HYDRANGEAS

HYDRANGEAS

beautiful varieties for home and garden

NAOMI SLADE

photography by

GEORGIANNA LANE

GIBBS SMITH
TO ENRICH AND INSPIRE HUMANKIND

Contents

INTRODUCTION

A FLOWER OF A THOUSAND FACETS, THE HYDRANGEA NEVER
CEASES TO SURPRISE AND ASTONISH. THEY ARE CHAMELEONS AND
SHAPE-SHIFTERS, MORPHING FROM FRESH AND VIBRANT YOUTH TO LANGUID
AND MYSTERIOUS AGE WITH NO DISCERNIBLE LOSS OF CHARM OR INTEREST.
AND WHILE THIS PLANT MAY NOT HAVE ALWAYS BEEN UNIVERSALLY LOVED,
IT CARES NOT ONE IOTA. THE HYDRANGEA IS HERE TO STAY; WITH US
ALWAYS AND FAMILIAR CERTAINLY, YET STILL IMBUED WITH GREATNESS.

Fashion is a capricious thing and hydrangeas, more than many plants, have had their low points as well as their triumphs. Discovered but not applauded, passed over in the annals of botanical significance, given away as an also-ran by those who might have cherished them. Yet hydrangeas have slowly surged, gradually building a reputation and a following; not catapulted to glory as a manufactured pop phenomenon, but gaining recognition the old way, through hard graft and reliability, like the band that plays working men's clubs and back-street dives, building up and burning slowly to finally become a national treasure.

Throughout their history, hydrangeas have tended to divide people. Some think they are marvellous in almost every way; others consider them an abomination. Even the tastemakers disagree. American magazine and television mogul Martha Stewart loves them; pop legend Madonna reputedly loathes them. And, until relatively recently, I would have been with Madonna all the way.

When I first met hydrangeas, they were bulky, dated landscape shrubs. They grew in a row under the window in my grandmother's coastal garden, the flowers vast lumpen mops of dull pink and mauve that my granddad called 'Queen Mother's Hats'. My grandmother's nickname was Queenie and the reference to the majestic headgear must have amused her no end.

Hydrangeas were simply not to my taste. My young self preferred dainty wildflowers, fragrant herbs and the juicy charms of the fruit cage. Yet it is unfair to judge an entire genus on a couple of neglected specimens viewed with an uncompromisingly critical pre-teen eye. Especially when said genus is going through a renaissance, a pop-star-style reinvention of the type where a dated crooner teams up with a hot young act and suddenly reveals that they can bust out the tunes in a whole new way, capturing hearts and exploding in popularity as they do so.

But youth is fickle and critical. It has a first taste of olives and they are the most revolting thing in the world; a glass of cheap plonk leads to a hotly declared dislike of red wine. There is, as yet, no appreciation of uniqueness, quality and subtlety, whether that be the olives from Manzanilla, Kalamata and Morocco – fruity, salty or tart – or the endless complexities of grape, climate and soil that go into crafting a fine wine.

So it is with hydrangeas. Reflection and experience, an appreciation of new developments and the simple turning of the world has made them not just freshly relevant but ultimately desirable. They are now courted, coveted and cooed over, wherever they can be grown or shipped to.

While the bulky old faithfuls still exist, they are a renewed force in a landscape or woodland garden design. And they are now joined by newer, more compact plants; plants that are ideal for containers. Breeders have developed fresh lacecaps, airy as a bridal veil, and elegant, sophisticated panicles in cream and green. They offer the exquisite excitement of a flower that ages not just gracefully, but magnificently, with antique shades of verdigris, teal and damson, before finally fading to a spare but deliciously delicate skeleton in the garden.

The versatility of hydrangeas must, in part, have contributed to their renaissance. They are perfect as a container specimen and house plant, and are suited to floristry of all kinds. In the garden, they are design magic. Hydrangeas as grand shrubs that flex their landscape muscle and compete with trees for impact; hydrangeas as modest-yet-handsome additions to a mixed border; those varieties whose large and lavish blooms thrust them into the spotlight as spectacular specimen plants; and those bred for neatness and good behavior, a gift to urban gardens and small spaces.

The consumer has become increasingly sophisticated, too, and there is a greater recognition that plump, bosomy mopheads are not the only option. The understated delicacy and variety of lacecap flowers, so much in the Japanese tradition, appeal to gardeners that prefer their plants to be more violin concerto than big band.

In the West, hydrangeas crept on to the gardening scene in the eighteenth century, some brought in from North America and others collected from the plethora of species to be found in Asian countries such as Japan, Korea and China. Growing wild and free, the *Hydrangea macrophylla* types comfortable in warmer maritime environments, while sundry hardier species extend into the mountains.

Here are remote populations of hydrangeas that have not made it into cultivation and may never do so. Yet, among them, are plants with horticulturally tempting qualities such as a towering arboreal form or a tendency to re-bloom. Such plants still evoke the tingle of excitement that must have been felt by the first explorers and collectors ever to penetrate these forbidden areas.

The Victorians, who were never backwards in coming forwards when attributing significance and whimsy, took a somewhat unflattering point of view when it came to hydrangeas, and seemed oddly unaware of the ironies of sending flowers to insult the receiver. A bunch of hydrangeas on your doorstep apparently implied that the sender thought you were a braggart (not that any popinjay worth their salt would care). A rejected suitor with wounded *amour propre* and an axe to grind might, similarly, send hydrangeas as a floral slap and accusation of frigidity – which would probably only help to cement the giver's position as "well and truly dumped."

In Asia, the relationship with hydrangeas is comparatively dignified yet ceremonial, and regionally quite complex. In Japan, at least, the hydrangea carries less horticultural significance than it does cultural significance. Other countries, meanwhile, seem relatively indifferent, although there is a flourishing market for the cut flowers in Hong Kong, China and Singapore. Across the board, the picture that emerges is as

emerges is as might be expected, as regards something that is a relatively common native flower, rather than an exotic garden specimen.

As the flowers evolve, they play out maturation, age and death, thus evoking change – or perhaps a fickleness or inconstancy. In Japan this mutable nature carries the implication of low status, yet hydrangeas can also be synonymous with grace and gratitude. More recently, it has been said that they represent love and true emotion, with meanings including "You are the beat of my heart"; a curious echo of the culturally distant Victorian "Language of Flowers."

In Asia, the hydrangea was traditionally a flower of condolence and, even now, some of the finest displays are to be found around temples and shrines.

In Japan it is also a focus of cultural celebration, in the same way as are cherry blossom and peonies. The Japanese for hydrangea is *ajisai* and as the flowers hit their peak in the rainy season (*tsuyu*), it is customary to visit collections of thousands of plants in full bloom in an act of ritual appreciation – an *ajisai* festival. Because the more moisture a hydrangea gets, the better it looks, prayers are offered up that the rains will not cease until the blooming is complete.

The plant is culturally significant in other ways, too. Like many other genera, hydrangeas contain a small amount of cyanide – indeed, they are often marked as toxic when sold commercially. Having said that, *H. serrata* also contains a sweet compound called phyllodulcin,

and the leaves are widely used to make a prestigious herbal tea. Known as *sugukcha* in Korea and *amacha* in Japan, the ceremonial infusion is used to celebrate the birth of Buddha, giving rise to its alternative names of "tea of heaven," or "Buddha tea."

In this book I explore this flower of mystery and rejoice in its evocative, absorbing, chameleon bloom. It is a contradictory plant that means so many different things to so many different people; lauded and cherished on the one hand, rejected and maligned on the other. Yet whether it emerges, luminous, from a shroud of summer mist on an Asian mountainside or graces a container in a modern home with its elegant simplicity, the hydrangea renaissance is now indisputable. Quietly, these flowers have crept into our interiors, gardens and hearts, while also invading the very bedrock of human culture.

As proof of our passion, thousands of hydrangea varieties now exist, while new ones are bred each year. Some will stand the test of time and others will prove ephemeral, but the genus *Hydrangea* will persist alongside us, giving rise to new traditions and perhaps even inspiring new myths. Blissfully indifferent to the vagaries of fashion, hydrangeas have taken center stage. And, as the performance continues, we cannot help but be seduced; swept up in the wave; enchanted by the floral siren song. Overwhelmed by a compulsion that leaves us in no doubt; certain by every possible measure that there is no situation in home nor garden that cannot be improved by adding a hydrangea.

THE HISTORY AND CULTIVATION OF HYDRANGEAS

THE STORY OF HYDRANGEAS SPANS THE WORLD. FROM THE OREGON WILDERNESS TO THE COOL, GREEN HAZE OF THE JAPANESE MOUNTAINS, IT IS A PLANT OF UNITY AND DIVISION. THE PROGENY OF POPULATIONS SEPARATED BY THE INEXORABLE, CREAKING DRIFT OF TECTONIC PLATES RIPPING CONTINENTS APART, YET REUNITED MILLIONS OF YEARS LATER IN OUR GARDENS. A PLANT THAT CREPT IN UNLAUDED, TO PREVAIL IN A BATTLE FOR POPULARITY THAT WAS NEVER OPENLY WAGED.

Hydrangeas are universally familiar plants. Embraced as shrubs and container specimens, they are lauded in the glossy pages of magazines and cooed over on social media. Yet despite their current prominence and popularity, hydrangeas didn't exactly arrive on to the horticultural scene with a bang. Rather, they crept in under the radar and prevailed by stealth. Theirs is a history set against an intriguing backdrop of colonialism, politics and exploration, trade and even diplomacy – or lack of it.

While they are now grown all over the world, wherever soil and climate permits, hydrangeas are actually native to a wide area of southern and eastern Asia, including China, Japan and Korea, parts of the Himalayas and Indonesia. An additional two species, *Hydrangea arborescens* and *H. quercifolia*, are found in the USA, and there are many more in Mexico and South America that have not made it into cultivation.

A number of plants, including magnolias, *Liquidambar*, viburnums and lilies share this divergence in geographical range, and it is thought that the disjunction is due to the separation of land masses by continental drift. Approximately 75 species of hydrangea have been identified thus far and, while Asia is home to the majority of these, there are dozens yet to be discovered, both there and in South America.

The hydrangea hunters

The first hydrangea to reach Western cultivation was the North American species, *H. arborescens*, which arrived in England in 1736, later to be named by renowned taxonomist Carolus Linnaeus. And, although it may not have made much of a splash on its own, it arose from the very crucible of American botany.

John Bartram was a man of his time. He was a colonial farmer and family man who worked hard in tough, pioneer conditions and did well because of it. He had no formal education, yet his interest in plants ran deep; he devoted a small patch of his farm to curious and medically useful specimens, and read whatever he could.

The young John was understandably cautious about travel: his father had been an evil-tempered man who, following his excommunication from the Quaker church, met a sticky end at the hands of Native Americans, but the lure of botany proved too much and over the years he explored the eastern colonies widely, travelling to Ohio, Ontario and Florida.

His big break came when he made contact with Royal Society member, merchant and fellow Quaker Peter Collinson, who commissioned him to send boxes of plants and seeds to London in a relationship that would last 30 years. "Bartram Boxes," as they became known, were instrumental in introducing a huge variety of North American plants to Europe and John himself was honored by Linnaeus as "the greatest natural botanist in the World."

A love of plants often runs in families, and John Bartram's third son, William, soon became a botanist and natural historian too, and together they discovered *H. quercifolia*, which remains a staple of American gardens today.

On the other side of the world, meanwhile, a different set of challenges faced Western plant hunters. Asia was a rich and exciting botanical seam, with an abundant flora just awaiting discovery, yet the gates remained firmly shut. Japan, in particular, was adamant that their society should not be contaminated by unsavory Western influences.

There was, therefore, only one legitimate way in: trade. The Dutch East India Company, a major player in global trading, was permitted a limited base – most countries, after all, are open to opportunities to become more wealthy – but that was restricted to a tiny, artificial island called Deshima in Nagasaki harbor.

With little space for personnel, it was usual for people on trading boats and bases to double up on skills. So doctors were often botanists, explorers, and much else besides, embodying that historically magnificent notion of "men of science." But, unfortunately, for Engelbert Kaempfer, physician and hunter of medically useful plants, when posted to Nagasaki with the Dutch East India Company, he discovered his botanizing was to be severely curtailed.

But the enterprising Kaempfer grabbed his rare opportunities with both hands and, on the 745-mile annual pilgrimage to pay respects to the Japanese emperor in what is now Tokyo, he discovered a number of plants. His haul included Asian hydrangeas, which were initially described as part of the *Sambucus* family, presumably as a result of the flat corymbs of lacecap flowers having a passing similarity to those of elderberries. He published his observations – *Japanese Flora* – as part of his travel writing, and his book *The History of Japan* followed, posthumously, in 1727.

The next hopeful naturalist to arrive on Asian shores was Carl Peter Thunberg, a Swede whose relatively brief medical study tour to Paris in the 1770s had developed epic mission-creep. From France, he went to South Africa, where he remained for three years before moving on to Japan. He, too, was based in Nagasaki and, although not allowed off the island, the "Linnaeus of Japan," as he became known, came up with a devious yet elegant ruse to acquire plant material. It was simple and admirably effective: he obtained a pet goat. And with a grazing

animal to provide for, it could only be reasonable for Thunberg's Japanese servants to go to the mainland to collect fodder.

The goat food proved a rich source of interest to man and beast alike, and yielded, amongst other things, two plants that he described as *Viburnum macrophyllum* and *V. serratum* – which were later re-allocated to the genus *Hydrangea*.

Then, in 1823, German-born Philipp Franz von Siebold arrived at the by now well-established trading post and promptly fell in love with all things Japanese – including his wife. An energetic and versatile man, his skills as an eye specialist gained him permission to visit the mainland, which proved to be a cornucopia of botanical temptation. He collected vigorously, setting up a glasshouse and garden at his home, and, unbeknownst to the Japanese authorities, sent home plants and seeds that included wisteria, hosta, magnolia and tea – which was a closely guarded commodity at the time. He is credited with the discovery and introduction of *H. paniculata* and *H. involucrata,* and he was also responsible for the introduction of Japanese knotweed, *Fallopia japonica*, an invasive menace, which has not been so universally welcomed.

But passion, chutzpah and opportunity proved too much, and when his boat ran aground and he was discovered with illicit maps of Japan, he was imprisoned as a spy. Expelled from the country in 1829, he returned only briefly, 30 years later, and spent an ecstatic three years as advisor on the West to the Japanese government before being diplomatically duped into leaving by his disenchanted former employer, the Dutch East India Company. And, sadly, he returned to Europe.

Thus by the mid-nineteenth century, hydrangeas were creeping inexorably into cultivation, and the next character on the scene was the somewhat hapless-sounding Charles Maries. He had joined the famous Veitch nursery in 1876 and, proving competent, was invited on an expedition to the Far East. His fate there was checkered. On the one hand, he collected and sent back a number of significant plants to his employers in England. On the other, however, he got sunstroke and then lost his seeds in a boat accident, which meant he had to re-collect them. And while he was apparently musical, which the native people liked, he also appears to have lacked personal charm. His refusal to try to understand and communicate with his local colleagues ultimately led to threats, robbery and the destruction of his collections.

In the end, he left for India, where he became an expert on mangoes, but not before he had brought back a form of *H. serrata* and a coastal lacecap hydrangea that is still available as *H. macrophylla* 'Mariesii'.

Unfortunately for Maries, the powers at the Veitch nursery were underwhelmed by his hard-won hydrangeas, but they did at least offer them to colleagues in Europe, who were significantly more enthusiastic. The French, in particular, were quickly hooked and, following their exhibition at the Société Nationale d'Horticulture de France in Paris in 1901, hydrangeas became the darling of a number of legendary French breeders.

In particular, they were eagerly embraced by the Lemoine brothers, who spotted the fertile flowers in mophead hydrangeas and bred from those. And, hot on their heels, came Messieurs Mouillère and Cayeux, whose mission was to breed the most fabulous hydrangeas for indoor decoration (and who bestowed upon them some magnificent and lengthy French names, such as Générale Vicomtesse de Vibraye and Madame Plumecocq).

HOW THE HYDRANGEA GOT ITS NAME

The name *Hydrangea* is derived from the Ancient Greek words *hydor*, meaning "water" (and from which comes the root-word *hydr-*, meaning "pertaining to water," as in "hydrant") and *angeion*, meaning a container such as a pitcher.

People love a good story and are quick to infer meaning, so it is sometimes stated that the name is an indication of the plants' thirsty tendencies and love of moist ground. It is even surmised that the name actually comes from Hydra, the snake-haired mythological monster, which the stamens could, with a modicum of imagination, be said to resemble.

But the real answer, or, at least, the most widely accepted one, is that the buds of the flower, before they burst, are the same shape as an ancient Greek vessel that was used to carry water.

Since then, plant breeders in Belgium, Switzerland, Holland and Germany have joined the race, with some fantastic new varieties coming from American and Asian breeders too.

As an historical footnote, it is interesting how this early appreciation of hydrangeas in Europe and lack of it in the UK has fed through to modern perceptions. Because most hydrangeas were originally *macrophylla* cultivars bred for the pot-plant market, they didn't always perform when transferred to the garden (see page 213 for advice on planting out hydrangeas) and so they got a reputation for not being hardy.

And because in the UK there is no tradition of breeding hydrangeas, there has been no commercially driven promotion on the scale that there has historically been in France and Germany, and now is in the US; as a result, they have been moderately ignored. Yet in America, the hydrangea basks in the limelight, described as "the plant of the millennium." And its reputational reparations have come full circle, as its adaptability as a long-flowering landscape shrub, together with greater awareness of the variety and

potential sophistication of the flowers, is putting the plants on the map, across the board.

Hydrangeas in cultivation

At time of writing, the *Royal Horticultural Society Plant Finder* lists 1,876 cultivated varieties of hydrangea. They are popular and versatile as garden plants and have been embraced for their lifestyle potential, too. In the UK, a hydrangea is – or was – most often to be found as part of a shrub border, or occasionally cut as a dried flower, but in the rest of Europe and America, the glamorous and long-lived flowers mean these shrubs are often sold as potted plants for indoor use.

As a result, there is a wide range of dwarf cultivars, designed to be grown as house plants. The cut stems are also in vogue with florists as statement flowers in bouquets and to make magnificent single-variety displays.

Despite the number of species and the global extent to which the genus has been adopted,

TYPICAL HYDRANGEA LEAF

HYDRANGEA QUERCIFOLIA

only a handful of hydrangea species could be considered common in cultivation. These are cultivars of *H. macrophylla*, *H. serrata*, *H. paniculata* and *H. quercifolia*, with *H. aspera* and climbing *H. petiolaris* less frequent, although far from unusual. There are also a number of hybrids, such as *H.* 'Preziosa' and *H.* Runaway Bride® 'Snow White', which combine the assets of parents of different species.

Generally, hydrangeas are thought of as shrubs, although they can range in size from a compact 12 inches or so to something that can resemble a small tree, and there are hefty climbers within the genus, too. In stem and leaf, they are morphologically quite plastic – the exact shade of leaf green or, when it comes to the stems, the degree of brown to taupe and level of spotting and bark-flaking can vary, and it is often necessary to wait for the flowers to appear in order to achieve a precise identification.

The leaves are generally oval in shape, with serrated edges and pointed tips – characteristics that are somewhat fluid, depending on the variety and, to some extent, the position and proclivities of the plant. The exception is *H. quercifolia*, which, as the name suggests, has leaves like an oak or elongated maple. And, while a number of hydrangeas have attractive autumn foliage color, this species creates a particularly good display.

Anatomy of a hydrangea

Flowers exist so that a plant can breed. In most flowers, the fertile organs of reproduction are at the center of the bloom, surrounded by delicate, brightly colored petals that are designed to attract the pollinating insects. In the bud, the whole flower package is protected by tough sepals, which open up and allow the flower to expand when the time is right.

When the pollen has been transferred from the male stamens to the stigma and ovaries, the flower starts to die back. The petals fade and fall, and the fruit or seed head forms.

In hydrangeas, however, the arrangement is slightly different. Here, the fertile flowers are tiny and insignificant, and the protective sepals that are usually green and relatively transient in other plants have evolved so they are large and colorful – even scented in some cases. It is these that attract the insects to the flower head and, although they have no reproductive organs and nothing in the way of reward to offer, once the insects are there, the true flowers will be pollinated.

Because of this, although the real flowers are transient in pursuit of their higher purpose of setting seed, the sterile florets, usually made up of four large sepals in the shape of a simple flower, have no such agenda, so can hang around looking good for months.

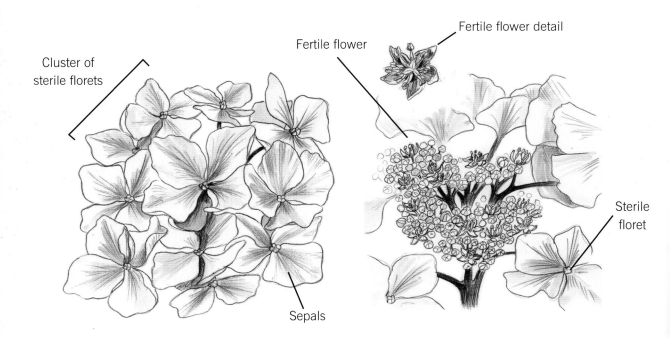

Cluster of sterile florets

Fertile flower

Fertile flower detail

Sterile floret

Sepals

STERILE FLOWERS

FERTILE FLOWERS

Flower forms

MOPHEAD

This hydrangea is the classic 'hortensia' type, which alternative name is used particularly in France and in other areas of Europe. It is purportedly named for its resemblance to a traditional floor mop; however, to my mind it does also rather resemble a mob cap, or mop hat – a Victorian servant's hat that was a large, round cover-all and jolly good for keeping dust out of your hair. Look up mop hat and draw your own conclusions.

Etymology aside, mopheads are spectacular snowballs of sterile sepals and contain almost no true flowers. Because there is nothing to pollinate and die, these flowers are as near to immortal as an annually produced plant-organ can ever be and they live on and on for months, fading only gradually.

LACECAP

Closer in appearance to wild hydrangeas, lacecaps have an airy, central corymb of small, fertile flowers surrounded by large, colorful "flowers." When the real flowers are pollinated, they go over, leaving just the outer ring. As a result, this flower type is less high-impact and lasts less long than mophead varieties – but they are very frequently scented.

Lacecap flowers also come with an interesting twist, quite literally in fact, as when the nights become cool and autumn advances, the large florets around the perimeter turn from gazing upwards towards the sky, to facing demurely downwards, as the flower senesces.

The name derives from the small frilly cap that perched on top of the neatly coiffed heads of maids in grand nineteenth-century homes (presumably those higher-status maids, who were to be seen by the guests so were not dressed for kitchen duties). In old pictures, the hydrangea-like effect of the lacy central cap surrounded by pillows of hair is remarkably apposite.

PANICLE

A panicle is an inflorescence, or flower head that has many branches along the stem. The species *H. paniculata* and *H. quercifolia* both produce panicles, but although the elegant, cone-shaped blooms can look quite solid, they, too, are made up of sterile ersatz flowers and clusters of small, fluffy true flowers.

MOPHEAD

LACECAP

PANICLE

Color chameleon

When the first Asian hydrangea reached maturity in The Royal Botanic Gardens, Kew, in the late 1700s, the flowers caused a sensation when their original color gradually changed – as if by magic.

Widely believed to be unique to hydrangeas, this chromatic instability is one of the plant's most memorable characteristics, and it comes down to the availability of aluminium ions in the soil.

Acid soils, i.e. those below pH7, contain aluminium ions in a form that the plant can absorb and then use to produce the pigment for blue blooms. Alkaline soils, which are above pH7, and those that are high in phosphate "lock up" the aluminium so the ions are unavailable to the plant. As a result, the same variety will produce flowers that are pink or mauve. The more acid your soil is, therefore, the more clear and intense the blue color will be.

Hydrangeas can take several years to settle into a site and assume their ultimate color, but giving them a dose of "free" aluminium in the form of potassium aluminium sulphate (potash alum) can speed up the process – but take care not to overdo it, as the plant may suffer.

If your soil is acid but you want to grow red hydrangeas, or it is alkaline and you want to grow blue hydrangeas, the simplest thing to do is to grow the plants in containers of appropriate compost. While it is possible to alter soil chemistry to a certain degree, the underlying characteristics of the soil, and those of the rocks upon which it lies, have considerable inertia and, frankly, it is not that easy to get good or fast results.

If you want to attempt it, adding lime to acid soils will sequester the aluminium ions and render them unavailable, but this needs to be done several times a year. Meanwhile "acidifying" alkaline soil is extremely difficult, if not impossible. Treating the ground around the plant with iron sulphate ($FeSO_4$) can free up the aluminium ions – but only temporarily.

When tinkering with soil pH and other factors that might affect color, it's important to be aware that a garden is a holistic system and a plant needs a range of minerals and micronutrients. So, by focusing on the availability or otherwise of aluminium, you may inadvertently reduce access to other nutrients, both for the cherished hydrangeas and for other plants, too.

Furthermore, the color-change phenomenon only works with hydrangeas that actually produce the appropriate pigment molecules – which are primarily the pink or blue forms of *H. macrophylla*, although *H. serrata* can be moderately susceptible.

The varieties that produce white flowers, therefore, such as *H. paniculata*, *H. arborescens* and *H. quercifolia*, together with the white forms of *H. macrophylla* and *H. serrata* are not affected by soil pH and are stable wherever they are grown – although they will often "antique" in autumn (see page 32). *H. aspera* will remain white.

It is also worth noting that many hydrangeas have a "preferred" color, and they will lean towards this, regardless. Simply, some would rather be pink and some would rather be blue, but their inherent tendencies are pushed one way or another by the soil pH. In some cases, this can result in some confusing outcomes, such as *H. serrata* 'Blue Deckle', which wants to be blue but, in very alkaline soil, can be pink.

Designing with hydrangeas

In the garden, hydrangeas are handsome and versatile shrubs. They excel in a woodland setting, particularly if you choose cultivars with lighter-colored flowers, and they can make a spectacular specimen in a mixed border.

Consider using them as an informal hedge or avenue. This treatment is perfect along the edge of a path or driveway and generates a sense of journey or momentous arrival. Alternatively, containers of matching plants, set out along a path or up a flight of steps, are elegant and formal.

Hydrangeas work well with complementary herbaceous plants, such as heleniums or *Hemerocallis*, and also with evergreen shrubs that have an opposing season of interest, such as azaleas or sweet box (*Sarcococca*). And, while in full floral spate the hydrangea will steal the show, in the depths of winter, the denuded shrub, with its charming, skeletal flowers, adds useful structure and interest to the garden. Underplant the shrubs with small bulbs, evergreen ferns and early-flowering plants such as hellebores for the best effect.

Hydrangeas as a cut flower

Long-lasting and glamorous, hydrangeas are hugely popular as a cut flower, whether arranged by themselves or in combination with other plants, flowers and foliage. The frothy blooms have a singular glamour to them and the colors can be as high-impact or as subtle as anyone could want.

CUTTING HYDRANGEAS IN THE GARDEN

Hydrangeas are extremely popular in floristry and the best time to cut them, as with all flowers, is first thing in the morning, before the sun has had a chance to warm the air and heat the plants. At this point the blooms will be fresh and the cells will be plump and full of water.

Take a clean bucket of cold water with you to the plant and cut nice, long stems straight into it. When you get back indoors, trim the bottoms off the stems under the water, cutting at an angle of about 45 degrees. This will ensure that no bubbles of air remain in the stems that could block water uptake. Leave in a cool and shady place to have a nice long drink, for at least a couple of hours. The leaves can be thirsty and draw the water away from the flowers, so strip some or all of the leaves from the stem. A layer of damp kitchen roll or a damp tea towel laid lightly over the flowers will create a moist atmosphere and limit water loss while they are recovering.

Charming as they are, hydrangeas do have a reputation for wilting easily; fortunately there are a number of tricks you can use to limit their sulking and also to revive them.

Any fresh young flower that has put on a lot of sappy growth risks wilting when it is cut and deprived of water from its roots. With large, often heavy heads, hydrangeas can also be physically weighty, which further impedes their ability to stay upright on a flaccid stem.

The first trick is simple – the flowers wilt when they are very soft and fresh, so don't cut them too soon. As the blooms open, they will harden up and be less susceptible to water loss. So choose the older flowers that are fully expanded with petals that look more papery. If cut into a bucket at this point, and treated well subsequently, they are less likely to wilt.

An alternative is to sear the stems by cutting as above, then dunking the end of each one into about 2 inches of boiling water for 30 seconds, before transferring them to a container of room-temperature tap water. This effectively kills the end of each stem, but it does mean that the small tubes that draw up the water will remain open.

When the hydrangeas have been arranged to your satisfaction, keep the vase and the water in it scrupulously clean, as this slows the growth of bacteria. A drop of bleach or a small spoonful of vinegar in the water also inhibits bacteria and will make the flowers last longer in warm weather. You can also extend vase life by changing the water and re-cutting the stems every few days.

CUTTING HYDRANGEAS FOR DRYING

As hydrangeas fade, they become newly beautiful, assuming desirable silvery shades or evolving a shot-silk palette of blue, pink or green as they dry. Even when fully brown, or skeletal, they are exquisite in autumn flower arrangements, combined with seasonal leaves and seed heads, such as alliums, fennel and honesty.

Once again, don't cut them too fresh – young flowers will wilt and won't hold their shape. But later in the season, as they harden up and become papery, they are far more resilient and can be harvested with impunity. Simply cut the mature bloom on a long stem, then arrange in a dry vase or hang upside down in a cool, dry place. The flowers will dehydrate but lose none of their beauty, and last for many months.

'ANTIQUING'

Summer brings fresh blooms, yet as the season moves on, the hydrangea flowers transform. Most will morph gently as they age, however the most dramatic changes often only occur as the nights start to get properly cold, and the days shorten in earnest.

Autumn is the time when the fresh, clear colors of youth mature to vintage hues – as the flower moves into its antique phase. White flowers grown in sunshine can often turn fiery red while, in shade, the same flower may assume a subtle shade of green. Blue can become wine, lilac turn to plum and red to mahogany, maroon or currant, while other shades take on an almost metallic bronze or pewter.

These muted colors of age add a level of sophistication and are a huge asset to the plant. While flowers left *in situ* will eventually fade to brown, if they are cut when antique but before they are frosted, this last vintage hurrah of color will beautifully persist.

HOW TO REVIVE WILTED HYDRANGEAS

If you discover that your flowers have wilted – perhaps they have reacted badly to a hot car journey home or the sun has come in the window and caught them unexpectedly – don't panic.

Take them out of their vase or wrapping, strip any leaves off and immerse the entire thing, flower heads and all, in room-temperature water. Leave them for several hours to rehydrate.

The shorter the stems are, the less far the water will have to travel to plump up the blooms, so when you rearrange them, cut the ends off and plunge them straight into the vase.

Societies and organizations

With widespread global popularity, it is no surprise that a range of societies, collections and gardens dedicated to hydrangeas have emerged. These are a great resource and your nearest society or garden should be your first port of call if you want to find out more, meet other hydrangea lovers and get tips and ideas for growing hydrangeas in your own area.

See www.hydrangeamania.com/id8.html for a list of interesting gardens.

United States of America
In the USA there are three independent hydrangea societies.

The American Hydrangea Society is located in the Atlanta, Georgia, area and covers the southeastern region of the country. Their website includes links to regional societies and other resources.
www.americanhydrangeasociety.org/about

The Mid-South Hydrangea Society is centered in Memphis, Tennessee, and serves the mid-southern states; it has a page on Facebook.

The Cape Cod Hydrangea Society is based in Massachusetts, where *Hydrangea macrophylla* thrives. In partnership with Heritage Museums and Gardens in Sandwich, they showcase one of the largest collections of *H. macrophylla* in the USA, celebrated with an annual hydrangea festival.
www.heritagemuseumsandgardens.org/cape-cod-hydrangea-garden
www.thecapecodhydrangeasociety.org

Canada
www.heritagehydrangeas.com is a website created by highly regarded hydrangea expert Barry Roberts, who is based in British Columbia.

United Kingdom
www.hydrangeaderby.co.uk A National Collection of *Hydrangea paniculata*
www.holehirdgardens.org.uk/plants/lakeland-collection-hydrangea

France
www.hortensias-hydrangea.com/hydrangea.anglais.htm
In their Normandy garden, Corinne and Robert Mallet have one of the finest hydrangea collections in the world, with a membership scheme and newsletter.

New Zealand
www.glynchurch.co.nz The garden of author and horticulturalist Glyn Church and his wife Gail.

Elegant
and Airy

Otaksa

This charming hydrangea is an old Japanese cultivar that dates from around the 1820s, and was named by Philipp Franz von Siebold after Kusumoto Taki, the woman he lived with as his wife during his time in Japan while employed as physician and scientist to the Dutch East India Company.

In his 2001 book, *Hydrangeas*, expert Glyn Church suggests that this variety may be a naturally occurring mophead *macrophylla*. But to Siebold, a romantic young botanist in a strange land of beauty and wonder, it must have been just the sort of discovery you'd want to name after the captivating local girl who had become the light of your life.

The large florets form loose balls; pointed petals overlap slightly to give a soft, ruffled effect. In autumn, the flowers fade to a fabulously delicate shade of green, and if cut and dried before the frosts get them, they hold their color beautifully.

Siebold called his wife "Otakusa" (which is thought to be derived from *O Taki san*), and the plant has entered cultivation under a contracted form of the name. The couple had a daughter, Kusumoto Ine, who also became a practicing doctor – thought to be the first Japanese woman to have received medical training at this level.

Hydrangea macrophylla 'Otaksa' is a coastal variety so is not particularly resistant to frost but will thrive in warmer locations and naturally lends itself to seaside gardens because the shiny, waxy coating on the leaves makes it tolerant to wind, sun and salt.

....

Hydrangea macrophylla 'Otaksa'
Height Up to 8 ft
Spread Up to 8 ft
Flower size Large to very large
Habit Large shrub
Color Delicate sky blue in acid soils and ice-cream pink in alkaline conditions, with underlying hints of green and cream
Preferred situation Light shade, not too cold
Hardiness USDA Zones 6–7
In the garden Best in a larger garden, where it will hold its own in a woodland setting or shrub border
As a cut flower Beautiful, delicate colors, perfect as a bouquet perhaps to celebrate a christening or baby shower; dries well if cut in its antique phase

Grayswood

This dainty and charming lacecap is the perfect antidote to the bulky blooms of the hortensia or to panicles of preposterous proportions.

Native to Japan and Korea, it would be naturally found in light woodland and, in the garden, it benefits from the shelter of trees and taller shrubs. It will also perform better if given a good, deep mulch around the roots at least annually. Although a bit miffy about too much sun and scorching wind, it is reasonably cold-hardy.

The leaves are oval, matt and neatly pointed, and the spotted stems sport loose, lacy flowers. Each of these has a coronet of nicely shaped larger blooms that have a central pinhead of smoky blue, within which is a loose central boss of blue-mauve true flowers.

Part of the allure of this plant comes from its chameleon transformation as the year progresses. Given a modicum of sun, the flowers evolve through a rainbow of crimsons and purples, while in deeper shade, it has a tendency towards greener hues.

The Royal Horticultural Society bestowed an Award of Garden Merit on Grayswood in 1993.

....

Hydrangea serrata 'Grayswood'
Height Up to 5 ft
Spread Up to 5 ft
Flower size Small to medium
Habit Small shrub
Color White bracts deepen to rose and wine shades around a heart of "true" flowers
Preferred situation Shade or partial shade
Hardiness USDA Zone 6
In the garden Works well in a woodland border or planted along a shady path
As a cut flower A delicate addition to a mixed posy

Grandiflora syn. Pee Gee

One of the showiest and most well-loved of hydrangeas, Grandiflora produces large panicles of sterile florets in summer, which mature through shades of apple blossom to a deep autumn pink. Before they fall, the leaves have a last flourish of yellow and gold.

The plant is popularly known as the Pee Gee hydrangea, which comes from the initials of 'paniculata Grandiflora' – "PG". The name literally describes the flowers: branched, conical inflorescences or panicles that are large in size – up to a hefty 18 inches long and 12 inches wide at their best.

This meaty plant is a dramatic addition to a large garden. The stems can be brittle so a sheltered site will prevent the wind damaging branches bowed under the weighty blooms.

As an elegant novelty, Grandiflora can be trained as a small standard tree with a single short trunk. The plant can then be pollarded, cutting the previous season's growth right back to a couple of buds. This both manages the size of the plant and reduces the number of flowers. The plant then puts all its energy into the remaining few and they grow vast and magnificent. It can also be grown as a bush or multi-stem shrub and left relatively unpruned. In this case there will be more flowers, but each will be smaller.

In the garden, these hydrangeas work well given plenty of elbow room. The habit is somewhat lax and sprawling, so young plants may need some support. Grow specimen plants surrounded with greens and glaucous hues – ferns and hostas are ideal, as are spring bulbs, which will fill the space with interest and color. Or plant for autumn color, with ornamental acers, *Acer campestre* (field maple) or *Cotinus coggygria*.

....

Hydrangea paniculata 'Grandiflora' syn. *H. p.* 'Pee Gee'
Height Can grow to 13 ft or more
Spread Up to 8 ft
Flower size Large
Habit Large shrub
Color Peppermint buds open into white bracts, deepening to rose pink and carmine
Preferred situation Sun or partial shade
Hardiness USDA Zone 3
In the garden Makes a great standard or specimen shrub
As a cut flower Make a statement by using as a single-variety bunch in a large, heavy-bottomed vase

Générale Vicomtesse de Vibraye

This decorative and charming plant was developed as part of the French drive to create ever more fabulous hydrangeas. Bred by Emile Mouillère, it is a hybrid of *H. m.* 'Otaksa' x *H. m.* 'Rosea' (not featured); introduced in 1909, it has been a firm favorite ever since.

Like all the colored macrophyllas, it is influenced by pH and, above pH7, Générale Vicomtesse de Vibraye presents itself an adequate pink, but it is on good acid soil that it reveals its true beauty. Here, buds of palest pistachio brighten and intensify, gradually taking on the cerulean tones of maturity. The pale green of the foliage only adds to the sense of freshness, and cutting the flowers is like harvesting an armful of summer sky.

In the garden, the clear blue flowers will illuminate a shady corner, looking fantastic with fresh whites or lemon yellows. Try planting with clumps of *Hemerocallis* or alongside simple, open, yellow roses. Alternatively, use them in a shrub border, under white-stemmed birches or with a plummy element to pick up the purple fleck in the stems.

The shrub itself grows into a moderate-sized spreading bush. It is highly floriferous and if the apical buds are lost to frost or grazing animals, the plant will flower readily from the lateral shoots. It will take full sun or partial shade, but the flowers can scorch in too bright and warm a spot, so give it plenty of moisture at root level. In drier areas, site it where it will receive some afternoon shade.

It was honored with a Royal Horticultural Society Award of Garden Merit in 1995.

....

Hydrangea macrophylla 'Générale Vicomtesse de Vibraye' syn. *H. hortensia* 'Générale Vicomtesse de Vibraye'
Height Up to 5 ft
Spread Up to 8 ft
Flower size Medium
Habit Medium shrub
Color Pale green buds deepen to pink on neutral soil and a beautiful blue on acid soil
Preferred situation Sun or partial shade
Hardiness USDA Zones 6–7
In the garden A dependable shrub; can be kept compact enough for a smaller garden
As a cut flower The sky-blue or pinkish flowers are very useful in bouquets as they are not too bulky; combine them with white flowers such as lilies and roses, or team them with stately delphiniums and lashings of acid-green *Alchemilla mollis*

Beauté Vendômoise

This elegant heritage variety does not scream and shout, but it attracts attention nevertheless.

Bred in France by Emile Mouillère in 1908, the handsome lacecap flowers have larger-than-average infertile florets: loose, airy, well-defined diamonds that hover over the neat central blooms like protective butterflies.

The sepals are fresh and light with just a flush of pink or blue, depending on the pH of the soil, and their bold form is enhanced by the denser and darker fertile flowers. It is also profusely floriferous and the green leaves are veiled with petals, like delicate Chantilly lace, for a subtly stunning effect. Beauté Vendômoise looks particularly lovely when grown in dappled shade and near white-stemmed birches.

Hydrangea macrophylla 'Beauté Vendômoise'
Height Up to 6 ft 6 in
Spread Up to 6 ft 6 in
Flower size Large
Habit A pillowy shrub that can get quite substantial
Color White flowers with a soft undertone of blue or pink
Preferred situation Partial or dappled shade; avoid frost pockets
Hardiness USDA Zones 6–7
In the garden A lacy specimen shrub
As a cut flower A frothy filler flower, subtle and elegant

Tuff Stuff syn. Cotton Candy, Blueberry Cheesecake

This cracking little shrub is a cultivar of *Hydrangea serrata*, a species heaven-sent to bring hope to gardens that have proven inhospitable to the more tender *H. macrophylla*.

The ancestral home of Tuff Stuff and its *serrata* brethren is the chilly mountains of Asia, so it is notably tolerant of low temperatures. It also has a winning habit of repeat-flowering – producing blooms on both new and old wood, and steaming ahead all summer.

The showy lacecap blooms have a striking semi-double appearance, and can range from pretty coral pink to a blue-purple, with a central eye of greenish-white, depending on the pH of the soil it is grown in.

An adaptable garden plant, it is compact and tidy enough to thrive in a decent-sized container while also holding its own near the front of an herbaceous or shrub border. It only really needs pruning when there is old wood to be thinned out and in temperate areas it will grow happily in both partial shade and full sun, but it will be grateful for afternoon shade in a really warm climate.

....

Hydrangea serrata 'Tuff Stuff'® syn. *H. s.* 'Cotton Candy', *H. s.* 'Blueberry Cheesecake'
Height Up to 36 in
Spread Up to 36 in
Flower size Small to medium
Habit Small, mounded, deciduous shrub
Color Rose pink or blue-mauve with a cream center
Preferred situation Will be happy in both partial shade and with a reasonable amount of sun
Hardiness USDA Zones 5–9
In the garden Works well in a woodland border and in gardens where *H. macrophylla* struggles
As a cut flower Busy and bold, it lends itself to country-cottage style arrangements

Wedding Gown syn. Doppio Bianco, Dancing Snow

This newer variety has almost everything you could ask for. Neat and compact, it is ideal for containers and for modern and smaller gardens. It cuts well for posies. The rounded leaves are a handsome summer green before turning gold in autumn.

But the flowers are the thing. And what flowers they are! Every stem bears a bloom that is lush and lavish, each a bouquet in its own right. Every bloom is a perfect, fully double star, opening gradually from green to chartreuse, and finally to pure white. As more and more florets unfold, they unite to create a dramatic white dome, at the same time exceedingly restrained and utterly fabulous.

And as a final bonus, it takes the legendary ability of hydrangeas to bloom over a long period and knocks it into a cocked hat. This is a plant that can be in full and exquisite flower from mid-spring to mid-autumn, which, let's face it, is longer than the duration of some marriages.

....

Hydrangea macrophylla 'Wedding Gown'® syn. *H. m.* 'Doppio Bianco', *H. m.* 'Dancing Snow'
Height Up to 40 in
Spread Up to 40 in
Flower size Medium
Habit A compact and rounded shrub
Color Glorious, fresh white
Preferred situation Partial or dappled shade, somewhere not too chilly
Hardiness USDA Zones 6–7
In the garden A good container specimen, it also works well in small gardens
As a cut flower Excellent and elegant; perfect as part of a bridal bouquet and makes a refined table arrangement

Beni-yama

A vision of elegance and subtlety, this Japanese lacecap hydrangea is almost achingly modest. The plant is small, at barely 3 feet tall, and the flowers are pared back, too.

The spare, chalky florets are poised around a smoke-blue or smoke-pink central cluster of fluffy fertile flowers that appear from midsummer. Pale at first, they gradually become freckled with raspberry, a color that bleeds and intensifies into richest sorbet. In autumn the foliage turns red, likewise.

The larger flowers can have either three or four petals when they appear, even on the same plant and in the same corymb, but this minor inconsistency doesn't detract from the sense of perfection.

While Beni-yama works well as part of a shrub or herbaceous border and is perfectly happy with moderate shade, the best autumn foliage color is produced when the plant sees some sun.

Beni-yama was honored with a Royal Horticultural Society Award of Garden Merit in 2012.

....

Hydrangea serrata 'Beni-yama'
Height Up to 40 in
Spread Up to 40 in
Flower size Small
Habit Small shrub
Color White bracts deepen to raspberry as the season progresses
Preferred situation Partial shade or full sun in good soil
Hardiness USDA Zone 6
In the garden Works well in a woodland border and small enough to use in a container
As a cut flower Best used as a filler flower or as a series of single sprays in eclectic vases

Twist-n-Shout

Bred by the same folk that produced Endless Summer® The Original (see page 157), Twist-n-Shout is hailed as the first reblooming lacecap *Hydrangea macrophylla*.

And what a mover and shaker it is. The flowers are sophisticated and airy and, depending on the amount of free aluminium ions available in the soil, can range from cornflower blue to rose pink, and all stages in between.

Twist-n-Shout is a hybrid that combines the repeat-blooming qualities of *H. macrophylla* 'Penny Mac' with the handsome characteristics and red stems of another good variety, *H. m.* 'Lady in Red' (not featured). As a result, the plant is loose but not large, with striking, dark red stems that contrast rather fabulously with the dark green leaves and the luminous flowers.

This hydrangea is easy to grow as long as it has good soil and plenty of moisture; its resistance to powdery mildew is also an asset. Grow Twist-n-Shout in partial shade as it will scorch if it receives too much sun.

Hydrangea macrophylla 'Twist-n-Shout'®
Height Up to 5 ft
Spread Up to 5 ft
Flower size Medium
Habit An airy, medium-sized shrub
Color Soft yet striking, in blue or pink
Preferred situation Partial or dappled shade
Hardiness USDA Zones 5–9
In the garden A reliable performer for a woodland garden and also thrives in a well-watered container
As a cut flower Good

Zaunkönig

The time comes to every gardener when they must embrace the inevitability of their climate and soil type, and if you are growing on chalk or limestone, then the pursuit of blue hydrangeas may be more of a frustration than a joy.

But if what you have is alkaline soil, and if what you want is a hydrangea that produces flowers of a splendid and reliable red, then Zaunkönig is the plant for you. To be fair, it is no slouch on acid soil either – in that situation the flowers are a good, strong purple – but it really wants to be red if it can be.

In close-up, the flower is really quite enchanting. The rich, pinkish-red sterile flowers have petals with frilly edges overlapping each other slightly like tiny windmills. These stand proud of the central boss of fertile florets that open from a tight pink bud into a lilac sea anemone, just awaiting a passing pollinator.

Zaunkönig is part of the robust Teller Series of hydrangeas that was bred in Switzerland and given German birds' names. In this case, the name means "wren," a minuscule but loud bird that is fairly common in the US, Canada, UK and Europe, and which builds tiny, round nests.

....

Hydrangea macrophylla 'Zaunkönig'
Height Up to 5 ft
Spread Up to 5 ft
Flower size Medium
Habit A rounded, fairly substantial shrub
Color Richly bloody flowers in alkaline soil, more purple in acid conditions
Preferred situation Partial or dappled shade in a mild spot
Hardiness USDA Zones 6–7
In the garden A healthy plant for a warmer climate; good in a woodland setting
As a cut flower Combine with autumn hues such as orange and red chrysanthemums, rosehips, seasonal foliage and crab apples

Hydrangea seemannii

If you find yourself to have been categorically bitten by the hydrangea bug, and the common or garden varieties no longer suffice, you might want to check out the more obscure *Hydrangea seemannii*.

A self-clinging climber with glossy, evergreen foliage, *H. seemannii* originates in Mexico and, as a result, does best in a moderate or warm climate.

The domed flower heads are untidy, with the larger white florets springing haphazardly out of a fluffy, greenish pompon of fertile blooms. Yet they are attractive in a relaxed and naturalistic sort of way, and they also smell good, especially in the evening.

While it will grow in shade and even get on well with quite low light levels, it is frost-tender and resents the temperature dropping below 23°F. It will also appreciate a sheltered spot, where the wind won't scorch the foliage. Growing it up a building will, to some extent, offer a degree of frost protection as the stone and brick absorbs warmth during the daytime like a storage heater. Alternatively, *H. seemannii* can be trained up a suitably robust tree, but it won't thrive with dry roots, so ensure that it has plenty of moisture, particularly while it is establishing.

....

Hydrangea seemannii
Height 26–39 ft
Spread 8–13 ft
Flower size Medium
Habit Climber
Color White
Preferred situation Likes a sheltered spot, happy in shade but will take a little sun, too
Hardiness USDA Zones 8–9
In the garden A hefty climber befitting a substantial structure
As a cut flower Too delicate, really, but you could experiment with using it as a filler

Hydrangea heteromalla

Known sometimes as the Himalayan hydrangea or the Chinese hydrangea for its origins, and sometimes as the woolly hydrangea for its furry leaves, *Hydrangea heteromalla* is a fairly variable species that, to the untrained eye, bears a passing resemblance to forms of *H. aspera*.

At its best, it can be a tall and elegant creature: an upright shrub or tall tree with huge, hairy, lance-shaped leaves carried on bright red petioles, or leaf stems. And it is the leaves that give rise to the species name, derived from the Greek *heteros* – meaning "different" – and *mallos*, which refers to the matted, woolly, tomentose underside of the leaf.

The blooms are individually fairly quiet: a coronet of pink and mauve florets surrounding a hub of true flowers that are milky white or green in color. But some particularly choice forms will unite to impress, producing great tiered candelabras of corymbs, the multiple flowers making up a mega-lacecap.

It is, by and large, healthy and hardy, but the forms with larger flowers and leaves can be tender and grow better in warmer climes. It is also intolerant of drought and prefers a plentiful supply of water to the roots, particularly in summer and if grown in full sun.

The variability of the plant in cultivation is because many are grown from seed rather than propagated vegetatively, so it pays to be selective in your choice.

....

Hydrangea heteromalla
Height Up to 16 ft 4 in
Spread 8 ft–16ft 4 in
Flower size Medium to very large
Habit Tall and elegant shrub or tree
Color Usually pinkish, white or green
Preferred situation Sun or partial shade
Hardiness USDA Zones Usually OK to 6 but some forms are more frost-tender
In the garden Best in large gardens and landscape planting
As a cut flower Not really the best

Pale and Interesting

Annabelle

While many glamorous hydrangeas are the products of breeders' efforts, Annabelle belies its film-star good looks with delightfully wholesome country origins. Discovered growing wild near the village of Anna in Ohio, for which the plant was named, this hydrangea is a real girl-next-door and as American as apple pie.

By hydrangea standards, the plant is not huge, but it is fabulously abundant. Flower heads of tightly clenched buds appear with a distinct chlorophyll hue but, as they open, they expand to form magnificent snowball flowers. The mature sepals retain the subtlest green tracery, which means that while the flowers shine in the sun, they don't glare; blending companionably into the border.

Native to America, *Hydrangea arborescens* is hardy enough to withstand most extremes, making Annabelle valuable for beginners and colder regions. While the stems strengthen with time, the flowers are heavy, so it is worth supporting the young plants at first.

To enjoy this deciduous plant at its naked best, site it where the skeletal blooms will catch the light and sparkle in the frost, perhaps with handsome umbellifer seed heads and alongside grasses such as *Miscanthus sinensis* and *Calamagrostis* x *acutifolia* 'Karl Foerster', which, similarly, will look elegantly faded well into winter. Underplant with spring-flowering bulbs such as snowdrops, aconites, glory of the snow and dwarf daffodils; moving on to summer companions with a vertical accent, such as delphiniums or lupins.

Annabelle received a Royal Horticultural Society Award of Garden Merit in 1993.

....

Hydrangea arborescens 'Annabelle'
Height 6 ft 6 in
Spread 6 ft 6 in
Flower size Very large
Habit Medium shrub, may need staking at first
Color Green-toned white, aging to darker green and lime
Preferred situation Sun or dappled shade
Hardiness USDA 3
In the garden Performs well in light woodland or in a mixed border
As a cut flower One of the best; lovely as a few stems in a vase or a stylishly simple hand-tied bouquet, and an outstanding dried flower, either in its green-toned winter livery or when the petals have dwindled to bare, taupe skeletons

ABOVE: ANNABELLE RIGHT: LANARTH WHITE 82

Lanarth White

A reliable and lovely shrub, Lanarth White produces airy, elegant corymbs of lacecap flowers in midsummer. And, at its floriferous peak, the flurry of flower heads covering the foliage gives the impression that the shrub is caught in a blizzard of large, static snowflakes.

Like other *macrophylla* types, it is sensitive to soil pH, so the central cluster of fertile florets can be cobalt blue to dusky pink in color, yet the neat coronet of larger flowers that surrounds this is reliably white.

While many hydrangeas quail at something or other in life, this one is relatively fearless. As such, it is a good choice for a seaside or exposed garden in poorer soil, and it will even tolerate full sun and a certain amount of dryness around the roots – although it would be a good idea to water it well while it is getting established.

Lanarth White was given an Award of Garden Merit by the Royal Horticultural Society in 1993.

Hydrangea macrophylla 'Lanarth White'
Height Up to 5 ft
Spread Up to 5 ft
Flower size Large for a lacecap variety
Habit Small to medium shrub
Color Clear white sterile flowers around a central dome of blue or pink-mauve
Preferred situation Will take full sun as well as partial shade
Hardiness USDA Zones 6–7
In the garden Flowers liberally and the blooms last well; good as part of a woodland garden and will often take a tricky spot
As a cut flower Forms a delicate frill when cut fresh, and dries really well, retaining its light coloring

Bluebird syn. Acuminata

A thoroughly graceful and elegant hydrangea, the flowers of Bluebird take a classic lacecap form; shapely and understated with large, clear sterile florets around the periphery of a woad-blue dome of fertile blooms.

The larger flowers are sometimes sparse, but their paucity serves only to emphasize their individual exquisiteness. As in a design, the space around something can be crucial, enhancing it and enabling the observer to fully appreciate the focal point. And should the plant be cruelly criticized for lack of impact, the fact that the flowers are scented should more than repair this deficiency.

The blue color is reasonably stable with variations in soil pH, too, although it can be more mauve or even pale pink on a very alkaline site. In autumn, the large sterile florets turn to face downwards and are infused with magenta and mulberry, complemented by the foliage, which turns copper-bronze towards the end of the season.

The shrub is compact enough to suit small gardens and will also do well in containers. Like all *serrata* cultivars, it is moderately cold-tolerant but does not find exposed or coastal sites conducive, nor is it a fan of full sun.

Bluebird has been honored with an Award of Garden Merit from the Royal Horticultural Society.

....

Hydrangea serrata 'Bluebird' syn. *H. s.* 'Acuminata'
Height Up to 4 ft
Spread Up to 4 ft
Flower size Medium
Habit Compact, small to medium bush
Color Clear blue sepals surrounding a darker central dome
Preferred situation Partial shade
Hardiness USDA Zone 6
In the garden Flowers from around early summer to mid-autumn and prefers a more shady spot
As a cut flower Adds a softly understated froth to an arrangement of seasonal blooms

Sandra

To thoroughly and finally dispel the image of hydrangeas as lumpen and dated, one only has to look at lovely Sandra.

Borne on top of a neat mound of greenery, the lacecaps are just achingly pretty – flawless little florets that could have been crafted from sugar paste as cake decorations – save that hydrangeas are, of course, not edible.

Each individual sepal is a perfect fan shape, defined and rimmed with color that bleeds ever so slightly into the pristine white center. The precise shade depends on soil pH: pink or red in alkaline soil or clear violet in acid conditions, but this detracts not one iota from the charm.

As the season progresses, the flower color intensifies, becoming richer, more dominant and altogether more robust in its autumn livery. The foliage is pleasing, too, emerging bronze and maturing to green before a final flourish of mahogany and plum.

Sandra is one of the Dutch Ladies Series of hydrangeas. Others include Sheila, a pinker flower that has crimped and pointed petals with a raspberry-stained rim. Sabrina is similar, but more coral pink, while Selina is washed with pink or mauve with a subtle white eye.

Overleaf: Veitchii

....

Hydrangea 'Sandra'
Height Up to 40 in
Spread Up to 40 in
Flower size Medium
Habit Makes a softly rounded bush
Color White and pink
Preferred situation Partial or dappled shade; avoid frost pockets
Hardiness USDA Zones 6–7
In the garden A charming addition to mixed and shrub borders, and a good choice for a hydrangea that is not brightly colored but not pure white either
As a cut flower Really pretty; experiment with including them in wedding or christening arrangements

Polestar

With an ultra-compact habit and light, foamy flowers, Polestar is a relatively new variety of hydrangea, and it has been winning hearts and minds since the day it arrived.

Its list of assets is impressive. There are the good looks – the pearly buds develop into tumbling tiers of cream flowers that are in full spate by high summer and which take on a pink blush that becomes richer and redder as the season rolls on. There is the fact that it starts flowering much earlier in the year than many of its *paniculata* brethren, in early summer or even late spring. And it is undemanding and easy to grow, so it is an excellent plant for time-poor gardeners.

But perhaps its most magnificent quality is its size. Polestar is a hydrangea that manages to be truly dwarf but loses none of its flower power in the process. So, in a world where there are more and more small gardens, terraces, exposed balconies and windowsills, where plants have to be robust performers that tolerate anything that life throws at them and still look good and work hard, Polestar is a fine choice indeed.

....

Hydrangea paniculata 'Polestar'®
Height Around 20 in at maturity
Spread 20 in
Flower size Small to medium
Habit Small shrub
Color White, becoming sugar-mouse pink with age
Preferred situation Sun or partial shade
Hardiness USDA Zone 3
In the garden Perfect for containers
As a cut flower Dries well and makes a nice frothy filler, but in many ways, why would you cut it when you can grow it on your windowsill?

Penny Mac

This magnificent mophead is named after hydrangea champion and queen Penny McHenry from Atlanta, Georgia, who, in 1994, founded the American Hydrangea Society.

And it is a star plant indeed. The glorious, juicy globes of flowers are produced repeatedly all summer, appearing the most beautiful azure blue on acid soil and taking on lavender tints in the presence of lime. And, as it wavers between the two states, the details can be glorious, with softly bleeding or marbled tones as the flower resists change and then relinquishes to its environment.

The remontant nature of this hydrangea effectively shortens the time it takes to produce new, flowering wood, and this means that Penny Mac is more tolerant of late frosts than many other cultivars. It is deemed more cold-tolerant, too, which makes it a good variety for gardeners in cooler regions – where macrophyllas are only marginally hardy – who are bold enough or foolhardy enough to want to experiment with this species.

While popular in America, it may be less easy to get hold of elsewhere, so a good alternative is Nikko Blue (see page 196).

Hydrangea macrophylla 'Penny Mac'
Height Up to 5 ft
Spread Up to 5 ft
Flower size Medium
Habit A rounded, medium-sized shrub
Color Ranges from clear blue to blush-pink
Preferred situation Partial or dappled shade
Hardiness USDA Zones 5–9
In the garden A robust specimen shrub
As a cut flower The cheerful globes suit substantial arrangements and work well dried

Maculata

It must be impossible to weary of hydrangeas but should the discerning gardener's palate ever get jaded, then Maculata is on hand to revive it.

In a group of plants that is, in most cases, all about the flowers, this is an elegant and energizing form with unusual variegated foliage. Thus, beneath large, white lacecaps, the pointed leaves of Maculata are rimmed and sliced with white and paler green, appearing more fresh and airy than the average *macrophylla* hydrangea.

This makes it a useful and striking garden plant – it is remarkable how such a simple change can make such a profound difference. If it has a flaw, it is that it is vulnerable to slugs and snails, especially when young, and they will slide over hot coals to devour it.

The cultivar itself is quite variable. The sparse coronet is comprised of florets that can have three, four, five or six component sepals, while at the center of the bloom is a bold cluster of fertile flowers, at first green but opening to become heliotrope blue or sometimes green-mauve or pink-mauve. The leaves, likewise, can be more or less variegated.

A number of other variegated hydrangeas are available, including the delightfully zingy Lemon Wave.

....

Hydrangea macrophylla 'Maculata'
Height Up to 40 in
Spread Up to 40 in
Flower size Medium to large
Habit Modest and airy shrub
Color White outer flowers with a blue or mauve center
Foliage Variegated green and white
Preferred situation Partial or dappled shade in a spot that does not get too cold
Hardiness USDA Zones 6–7
In the garden An unusual specimen shrub
As a cut flower Moderately good; charming rather than impactful

Phantom

While this handsome hydrangea may be of an eldritch hue, and while it may hover in the border and woodland, panicles upright and quivering like a collection of restless shades, its impact is far more stylish and dramatic "Phantom of the Opera" than alarming and ghastly "Dawn of the Dead."

The huge, dense, triangular panicles apparate subtly. The flowers are, at first, a vision of palest ectoplasm green, which gradually become paler, fading to white as the flower expands to become one of the largest of the paniculatas, at up to 15 inches long. Finally, the densely packed sterile florets fade into light pink as autumn signals a lingering end (see overleaf).

Despite their size, the flowers are held erect on strong stems, floating rather than flopping, which makes this hydrangea perfect in the shrub border against a foil of greens.

Hydrangea paniculata cultivars are deservedly popular and Phantom is easy to grow and tolerant of all soil types. It was given a Royal Horticultural Society Award of Garden Merit in 2008.

....

Hydrangea paniculata 'Phantom'
Height 6 ft 6 in
Spread Up to 6 ft 6 in
Flower size Large
Habit Large shrub
Color Delicate green buds open into white florets that become touched with pink
Preferred situation Sun or partial shade
Hardiness USDA Zone 3
In the garden Makes a great standard or specimen shrub
As a cut flower Spectacular but potentially impractically large

Star Gazer syn. Kompeito

For those who have a passion for the exotic things in life, who hanker for something out of the ordinary and who favor frills and furbelows, Star Gazer is a must-have plant.

Its picotee-edged flowers are fully double, a lacy frill of stars nestling inside stars, which gives each floret a layered, thee-dimensional quality, like the upturned tutu of a ballerina engaged in some unusual inverted maneuver. The central boss of pink or blue pearls, meanwhile, matures to a pleasing, complementary froth of stamens.

Being a *macrophylla* hydrangea, the color can range from sky blue to candyfloss pink and the flowers are borne on strong stems that keep them proud, even after rain. Like many of the more modern hydrangeas, it has been bred to repeat-bloom so it has a long flowering period together with a good resistance to mildew.

....

Hydrangea macrophylla Star Gazer syn. *H. m.* 'Kompeito'
Height 3–4 ft
Spread 3–5 ft
Flower size Medium
Habit A slightly spreading, medium shrub
Color Clear pink or blue, depending on soil type
Preferred situation Partial shade or dappled light under trees
Hardiness USDA Zones 6–7
In the garden Unusual and rather glamorous
As a cut flower Lovely and delicate; try with rosebuds and lilies for a formal bouquet or experiment with flowers and foliage of a steel-blue tone, such as eryngiums or eucalyptus

Magical Candle

Vigorous, robust and with its columnar, waxy blooms held aloft on strong stems, the really rather mighty Magical Candle absolutely declines to flicker feebly in the gloom when it can flame like a torch in the sunshine.

A cultivar not dissimilar to Limelight, it has juicy, densely packed panicles that are an unearthly yellow-green in youth, and that become larger, whiter and brighter as the summer progresses and they reach their dazzling peak. Finally, they start to spot and freckle with pink as the days begin to shorten and the flowers capitulate to inevitability and a graceful old age.

As with so many paniculatas, there is a real drama about Magical Candle and it is a good subject to show off in bright or exposed spaces, such as in a sunny front garden, planted as a hedge around a lawn or as a focal point surrounded by contrasting foliage.

Magical® is a US-registered trademark of Plants Nouveau, and Magical Candle is part of the Magical® Series, which includes the more strongly colored Magical Fire and Magical Flame (the autumn colors of this variety are shown overleaf).

....

Hydrangea paniculata 'Magical® Candle'
Height 3–4 ft
Spread 4 ft
Flower size Large
Habit Modest, slightly spreading shrub
Color Pistachio buds and young flowers become white and, finally, pink
Preferred situation Sun or partial shade
Hardiness USDA Zone 3
In the garden Makes a great standard or specimen shrub
As a cut flower A dramatic and unusual green element in arrangements, especially when cut relatively young

Blue Wave syn. Mariesii Perfecta

An established cultivar and in many ways unbeaten, Blue Wave is a very useful landscape hydrangea.

The flat, lacy flowers are produced from midsummer onwards and respond to the level of acidity of the soil; this means that despite the assumption of eponymous blueness, they can also appear in gentle shades of lavender, lilac and blackberry-ice.

In the garden, Blue Wave grows to become a large bush that is wider than it is tall. In full flower and on a good, acid soil, the overarching impression is of a green and sapphire explosion. If you apply a modicum of imagination, it really does resemble a wave flecked with spume and at the point of breaking, rushing towards the onlooker.

Blue Wave is not really suitable for small gardens but makes a wonderful display at the back of the border and in a woodland setting.

....

Hydrangea macrophylla Blue Wave syn. 'Mariesii Perfecta'
Height Up to 6 ft 6 in
Spread Up to 8 ft
Flower size Medium to large
Habit A rounded, medium-sized shrub
Color Blue or pink, depending on soil pH
Preferred situation Partial or dappled shade
Hardiness USDA Zones 6–7
In the garden A large, rather spreading shrub
As a cut flower Pleasant, frothy filler but it may drop

Pee Wee

This gorgeous miniature hydrangea may be compact but it has all the interest and refinement of larger plants, which makes it perfect for awkward, small spaces. A quart in a pint pot, it is also tough and hardworking – keep the water levels reasonable and give it some sun, and it will perform for months.

Pee Wee has dainty, oak-shaped leaves, which are matt and slightly leathery in texture; they color prettily in autumn – the more sun it gets, the brighter the hue will be. The panicles are similarly small and tight but the sterile white flowers are beautifully defined: minimal and fuss-free as if a cloud of white butterflies had landed on a buddleia bloom.

If you have a little bit of room, then Sike's Dwarf is a slightly larger alternative, but if your gardening is confined to a light well or a balcony, add a pot or two of Pee Wee to your collection and reap the rewards.

....

Hydrangea quercifolia 'Pee Wee'
Height Up to 40 in
Spread Up to 40 in
Flower size Small
Habit Neat, rounded shrub
Color Small, white panicles gradually develop a hint of pink
Preferred situation Partial shade or full sun
Hardiness USDA Zone 5
In the garden Provides flower and foliage interest over a long period and is ideal for smaller spaces
As a cut flower Airy and smaller than some hydrangeas, it is well adapted to floristry and celebration arrangements

Hydrangea aspera

Hydrangea aspera is less well-known than the macrophyllas, paniculatas and serratas, but it is handsome and garden-worthy, nevertheless.

Because of the huge extent of its natural range – it is found from the Himalayas and Burma, through southern and central China, and even in Asian island nations, including Sumatra – it is hugely variable. Indeed, many of the plants now classified as subspecies of *H. aspera* were once considered species in their own right.

So, to a certain extent, it defies definition and the wild-type plants can range from tiny forms of *H. a.* subsp. *strigosa* at less than 3 feet tall, to veritable giants of 33 feet in the favorable, warm and moist forests of the East.

For garden purposes, the flowers have a delicacy and understatedness that the more brash macrophyllas lack, and the foliage has a tactile quality all of its own. At best, the leaves are velvety and rich to the touch and create a beautiful green-plush foil to the blooms.

There are a number of good garden cultivars. Hardy *H. a.* 'Macrophylla' has a Royal Horticultural Society Award of Garden Merit and grows to a height of 6 feet 6 inches. *H. a.* (Villosa Group) 'Velvet and Lace' (AGM) is a superior plant of about the same size, while excellent, somewhat smaller cultivars include *H. a.* 'Peter Chappell' (AGM), *H. a.* 'Mauvette' and *H. a.* 'Farrell'.

....

Hydrangea aspera
Height Variable
Spread Variable
Flower size Small to medium
Habit Highly variable
Color Usually purple or pink and white
Preferred situation Benefits from a woodland-edge setting but not averse to a bit of sun; avoid frost pockets
Hardiness Very variable; seek out named cultivars and those with a reliable pedigree
In the garden Versatile and interesting as part of a woodland garden or group of specimens
As a cut flower Modest and not always very long-lasting

Cool
and Crazy

Vanille Fraise syn. Vanilla Strawberry

One of the most frothily joyful of modern hydrangeas, Vanille Fraise is a chameleon in floral form. Each small floret starts life as the safest of ice-cream hues before morphing gaily into a symphony of crushed berry and finally, into the dark and jammy realms of currant crumble. Viewed *en masse*, the pendulous, pink, cream-tipped panicles have a fantastical quality that would not look out of place in *The Cat in the Hat*.

The slightly lax habit of the plant makes it ideal for a cottage garden or informal setting. Large, heavy, conical flowers tumble out of a medium-sized shrub, and the fresh green leaves and reddish stems are a handsome complement to the display.

The flowers are abundant and can be cut fresh, to be used, perhaps, in a substantial arrangement with long stems of plummy *Physocarpus opulifolius* 'Diabolo', fluffy-headed grasses and *Verbena bonariensis*. They are also good when dried and combined with autumn leaves and twigs.

Vanille Fraise will grow best in moist, fairly fertile soil in light shade or partial sunshine, but its ability to thrive even in a shaded site makes it a useful garden plant. In fact, it is very hardy, surviving in temperatures as low as -4°F and, even if the stems are damaged by frost, the plant will still produce flowers on the new season's growth. In warmer regions it is best planted where it will receive shade in the afternoon, if not longer, and don't let it dry out.

Hydrangea paniculata Vanille Fraise was voted top plant of 2010 by the American Nursery and Landscape Association.

....

Hydrangea paniculata 'Vanille Fraise'® syn. *H. p.* 'Vanilla Strawberry'
Height 6 ft 6 in
Spread 4 ft
Flower size Very large
Habit Medium shrub
Color Creamy white, taking on pink tones that intensify as the flower ages
Preferred situation Partial shade
Hardiness USDA Zone 3
In the garden Works well in small and urban gardens, and in a low-maintenance border
As a cut flower The large, loose flowers look lovely arranged informally in a bucket or decent-sized vase

Limelight syn. Zwijnenburg

Shamelessly fuelling the current passion for hydrangeas, Limelight is a creature of curious form and intriguing hue.

Tightly virescent buds break in series to form a dense conical flower of eldritch acid green, the flowers becoming fresher and lighter as they expand. As the flower ages, it broadens and flattens, taking on a wash of deep pink, which intensifies until full maturity. Usefully, the flower color in this variety is stable, regardless of soil pH.

Resembling lavish, peaked dollops of pistachio ice cream, the flowers are arrayed tipsily over the greenery. The large, long panicles are creatures of substance and solidity, while the plant itself is upright and spreading, echoing the floral palette with its stems of greenish buff and fawn, and its leaves with a slight hint of citrus.

Although at first Limelight deports itself with muted elegance, its commitment to garden drama is revealed as the year wears on. The flowers develop in different stages of color and maturity, so young green-and-white flowers and older, pinker, ones are carried at the same time in an unusual three-tone effect. Finally, in autumn, it throws caution to the winds and the oval leaves follow suit, coloring brightly as a final hurrah before they fall.

Limelight is a great accent plant and, because it flowers on the current season's growth, it will continue to give a reliable display even if the stems have been damaged by frost.

Limelight has held a Royal Horticultural Society Award of Garden Merit since 2008.

....

Hydrangea paniculata Limelight syn. *H. p.* 'Zwijnenburg'
Height Up to 8 ft
Spread Up to 8 ft
Flower size Large
Habit Medium to large shrub
Color Green, becoming creamy with age, and finally taking on a pink flush
Preferred situation Full sun or partial shade
Hardiness USDA Zone 3
In the garden A striking back-of-the border plant; works well in small and urban gardens, and as part of a low-maintenance border
As a cut flower The unusual green flowers look good teamed with purple cactus dahlias, dark pink roses and silvery grasses; they also dry very well

Snowflake

Some plants have all the luck and, with exciting, pointed flowers and fabulous leaf color, this hydrangea is a double threat to those in its class.

Hydrangea quercifolia is native to the south-east of America and it takes its species name from the lobed leaves, which resemble those of an oak – *Quercus*. Although it still likes rich living and plenty of soil moisture, it produces its best autumn colors when grown in sun, which ripens the leaves to a blaze of purple, red and orange.

Double flowers are often bouffant, but Snowflake takes this even further. Each of the bracts lies over the one below in a series of spirals, building up a sumptuous and lavish-looking bloom that has a kind of stacked quality and considerable garden impact.

This plant tolerates both acid and alkaline soil and, in this species, color is indifferent to pH so the flowers remain a reliable white, turning pink as they age. Snowflake prefers a warm site and received a Royal Horticultural Society Award of Garden Merit in 2012.

....

Hydrangea quercifolia 'Snowflake'
Height Up to 6 ft 6 in
Spread Up to 6 ft 6 in
Flower size Medium
Habit Medium to large shrub
Color White double flowers turn rose pink in autumn
Preferred situation Partial shade or full sun
Hardiness USDA Zone 5
In the garden Provides flower and foliage interest over a long period and can be grown as a specimen shrub or in a mixed border
As a cut flower Dries well and retains pink hues when cut in early autumn

Glam Rock syn. Pistachio

In recent times, hydrangeas have been lauded as an elegant and sophisticated option in the modern garden or muted bouquet. But then, sliding out of the wings in a gold lamé jumpsuit and war paint, to tumultuous applause, comes Glam Rock.

This is a flower that doesn't care what you think and deports itself accordingly. The buds each open with an assertive shade of green and then they hit you with a pout of pink before embarking on an exhilarating crescendo of purple and blue at the core. Eye-catching and borderline-psychedelic, it is, like all the best rockers, a small plant but one that lives its life as a stadium event.

A master of reinvention, this magical, morphing floral freakout goes by several names. Pistachio suits it very well in its early phases but Glam Rock does it more justice later in the year.

This hydrangea was discovered as a chance seedling and was introduced in 2012 by Star® Roses and Plants. The tones of the relatively small mophead flowers will vary according to the soil in which it is grown, but that is par for the course and, given the nature of the beast, is probably best viewed as another interesting quality. Grow Glam Rock in a warm and sheltered spot in reliably moist soil, or let it put on a show in a container.

....

Hydrangea macrophylla 'Glam Rock' syn. *H. m.* 'Pistachio'®
Height 24–36 in
Spread 36–45 in
Flower size Small
Habit Small to medium shrub
Color Vivid green flowers mature to a tricolored combination of pink, green and blue
Preferred situation Sun or partial shade
Hardiness USDA Zone 6–7
In the garden Provides flower and foliage interest over a long period; an excellent specimen plant or container subject
As a cut flower A striking addition to an arrangement or great on its own

Runaway Bride Snow White

Rocking the hydrangea world at time of writing is a new hybrid, *H.* Runaway Bride®
'Snow White', which was voted Plant of the Year in 2018 at the RHS Chelsea Flower
Show, when it was introduced by seed company Thompson & Morgan.

This marvel is the work of Japanese plant breeder Ushio Sakazaki, who, having creating
many bedding plants, including Surfinia® petunias, turned his attention to hydrangeas.
The story goes that he came across a remote Asian species in the wild and, seeing
its potential, crossed it with common *Hydrangea macrophylla.* The resulting plant not
only produces blooms at the tips of the stems, like most other hydrangeas, but it also
produces flowers from every leaf node below.

The elegant lacecap flowers are wispy as meringue and smooth as icing, and they are
produced in sequence from late spring until autumn. The effect is airy and graceful, the
modest green shrub adorned with pearls and strewn with confetti; a vision of purity that
starts off a fresh, green-tinted white, and blushes to pink as maturity takes hold.

The plant itself is neat, compact and hardy. It looks great in a mixed border and, with
each garland of flowers cascading and tumbling, it is striking as a container specimen or
arching over the top of a low wall.

....

Hydrangea Runaway Bride® 'Snow White'
Height Up to 4 ft
Spread Up to 4 ft
Flower size Medium
Habit Small to medium shrub
Color Fresh white flowers gradually assume a rosy blush
Preferred situation Partial shade or full sun
Hardiness USDA Zones 6–7
In the garden Reliably remontant, it flowers over a long period; grow in pots or use as a
specimen shrub
As a cut flower A frilly, frothy addition to a bouquet; lovely by itself and equally pretty
combined with blues and mauves, or in combination with soft green foliage such as
asparagus fern

Preziosa

One of the most admired and adored hydrangeas around, Preziosa confuddles and charms with its rather singular approach to color.

Green flower buds expand through chartreuse and cream before blooming white. But this state of purity lasts only for the blink of an eye. Almost as you watch, a deep raspberry stain begins to bleed and splash the delicately frilled petals. The juicy hue seeps into the flower, like berry jam sinking into a slice of bread in a bowl of summer pudding; smooth gradients of color interrupted by more intense spots and splotches. Over the months, the Jackson-Pollock petals continue to morph, their underlying shade deepening to maroon before finally reaching its rich, winey, autumnal finale.

Although now classed (by the RHS at least) as a hybrid, Preziosa has a hefty slug of *serrata* blood in its veins, as the dramatic color changes attest. The white-to-red shift is stable and unaffected by soil pH – insofar as not ever turning blue is concerned, at any rate.

In the garden, this upright shrub prefers wetter soil conditions and at least some shade. In too warm or dry a spot it will scorch, so mulch it in spring to help retain soil moisture. It excels as part of a party, in a plant-packed border, cottage garden, or informally designed shady site. It is compact enough for smaller gardens, as well.

And while failing to remain a decently minimal number of colors may be too much for purists, for others Preziosa is a gem that can provide months of evolving garden interest.

....

Hydrangea 'Preziosa' syn. *H. serrata* 'Preziosa'
Height Up to 5 ft
Spread Up to 5 ft
Flower size Fairly small
Habit An upright, medium-sized shrub
Color The flowers start fresh white, morphing through raspberry to burgundy wine
Preferred situation Partial shade
Hardiness USDA Zone 6
In the garden Provides interest over a long period; tolerates a moist or shady site
As a cut flower There is so much going on with these flowers that they almost don't need any accompaniment, but if you do choose to combine them, pick colors that are on the white-pink-plum continuum, with perhaps a little hint of silver, such as roses, cosmos and cleomes, together with the seed heads of plants such as honesty (*Lunaria annua*)

Ayesha syn. Silver Slipper

Mophead hydrangeas can sometimes look a little bit overblown and ostentatious, but not so *Hydrangea macrophylla* 'Ayesha'. Each crinkled mound is made up of hundreds of small, uneven flowers with boldly concave petals, giving an effect which is charming and rather chic.

The impression is distinctive – it is clearly a hydrangea and, while not exactly lacy, it has an airier quality than the average hortensia. Its bobbly, textured appearance has led to its being referred to in some quarters as the popcorn hydrangea or the lilac hydrangea, and it carries a light but pleasant fragrance.

Highly susceptible to soil pH, Ayesha can range from brightest baby blue in acid soil, through lilac to white and sugar-mouse pink in alkaline soil. The petals are fleshy and individually variable in size and color, and infant blooms are prettily stained with pistachio green.

With bold, glossy, resilient leaves, this a good shrub for a more exposed spot or coastal garden, although it is more frost-tender than some. The leaves color attractively in autumn and the flowers, likewise, deepen in hue as the season progresses.

While Ayesha looks great in the border, the distinctive flowers lend themselves to being grown in a position of greater prominence, such as in a decent-sized patio container.

....

Hydrangea macrophylla 'Ayesha' syn. *H. m.* 'Silver Slipper'
Height Up to 5 ft
Spread Up to 5 ft
Flower size Small to medium
Habit Medium shrub
Color Variable, ranging from shell pink to bright blue
Hardiness USDA Zones 6–7
In the garden Flowers over a long period; grow in pots or use as a specimen shrub
As a cut flower Glorious as a single-variety arrangement and small enough to include in a hand-tied bouquet; try in combination with phlox, cosmos and the smaller single dahlias

Incrediball syn. Strong Annabelle

A modern and dramatically bouncy-looking shrub, Incrediball gives the impression of vigor and robustness before you even see it. The name alone speaks of a superhero-plant that is tough in the face of adversity, facing down evil pests and diseases with an expression that suggests the villains can try, but they'll never be cool enough or hard enough to prevail. A plant that is probably about to star in its own animated movie.

And, to a certain degree, first impressions do not mislead. Known also as Strong Annabelle, it lacks the original's tendency to flop in its early years, producing branches that are equal to the task of supporting the enormous mops of flowers, even after rain.

The flowers are magnificent, huge, dramatic globes of white, which start off large and increase in size as the plant matures. The flowers are not affected by soil pH and the green-washed buds open white, aging to jade at the end of the season. And if white is too vanilla, try its recently introduced pink cousin, Incrediball Blush.

Incrediball accepts sun or partial shade, but the hotter and dryer the climate, the more shade it needs. It is also hardier than *macrophylla* types and it is claimed that in a suitable microclimate, it will survive even in USDA Zone 3 – which drops to well below -30°F. It is also easy to care for; cut the stems back by about a third in late winter for flowers in late summer and early autumn.

In the garden, Incrediball is neither shy nor retiring, and benefits from being given plenty of space in which to show off. It excels as a landscape shrub or on the edge of a woodland garden, or combine it with large-leaved plants such as hostas, ferns or even hardy exotics.

....

Hydrangea arborescens 'Incrediball'® syn. *H. a.* 'Strong Annabelle'
Height Up to 6 ft 6 in
Spread Up to 6 ft 6 in
Flower size Very large
Habit A beefy, rounded shrub that needs plenty of space
Color Fresh white flowers aging to green
Preferred situation Partial shade or full sun
Hardiness USDA Zone 3
In the garden Makes a dramatic specimen or an interesting deciduous screening hedge
As a cut flower Arrange fresh in a suitably large vase, or cut and dry the blooms

Dark Angel Red

Unbelievably striking, Dark Angel Red is part of the new, award-winning Black Diamonds® Series of hydrangeas, which look set to become modern classics.

The textured leaves are large and the dark green is imbued with purple and near-black, and threaded with a filigree of lighter green veins. As autumn approaches, they color before they fall, smoldering redly like the hot embers of hellfire.

Atop a compact bush, the bold, flat flower heads appear in midsummer and consists of a tightly clenched caviar of fertile buds surrounded by striking florets. The center of each young floret is daubed with a creamy dollop of hollandaise sauce, while the tip of the petals look rather as if they have been dipped in strawberry coulis or stained with pickled beetroot, with the pink color becoming more dominant as the flower matures.

The color varies somewhat with soil pH, so you will get the best reds if you are gardening on chalk. But for those on acid soil, the blue-flowered sister cultivar Shining Angel will be no disappointment.

These hydrangeas are relatively new on the scene so it is worth being a little conservative when it comes to cold spots, but with their striking good looks and overwhelming garden presence, these angels are a force to be reckoned with.

....

Hydrangea macrophylla 'Dark Angel Red'
Height 20–32 in
Spread 20–40 in
Flower size Medium
Habit A rounded, spreading bush
Color Glowing crimson-wine flowers that deepen with age
Preferred situation Partial or dappled shade in a spot that does not get too cold
Hardiness USDA Zones 6–7
In the garden Striking at the front of a border and dramatic in a container
As a cut flower Attractive when cut and should also dry well

Spike

The ruffled, delicate blooms of Spike contradict long-held notions of stiffness or old-fashionedness in the genus *Hydrangea*, with their relaxed and contemporary style.

The flowers are textured and soft-looking, tactile rather than starchy and, depending on soil, they morph gently from summer-evening blue to dawn-pink, with celestial undertones of green to stop the confection from becoming too sugary.

In the garden or as a cut flower, Spike has a slightly blowsy quality, like a tight posy of hand-tied, old-fashioned Spencer Mix sweet peas, and can be combined prettily with scented companions if fragrance is a factor. Try cutting the flowers as single stems and displaying them in mix-and-match vases.

The flowers last well, too, providing color and interest all summer long before gradually antiquing to richer, deeper greens.

Spike was awarded a Silver Medal at the 2011 Plantarium trade show.

....

Hydrangea macrophylla 'Spike'
Height Up to 4 ft
Spread Up to 4 ft
Flower size Medium
Habit A rounded, medium-sized shrub
Color Gentle, pastel hues
Preferred situation Partial or dappled shade
Hardiness USDA Zones 6–7
In the garden A surprisingly tactile-looking plant for near doors and dining areas
As a cut flower Excellent in summer and autumn bouquets; the flowers are also good dried

Cocktail

Lavish, luscious and ever-so-slightly decadent, it is hard to know whether Cocktail is named for that deceptively innocent, creamy, fruity drink that packs such an unexpected punch, or for the glamorous frock that one would wear to a soirée where such beverages would be consumed.

The serrated foliage is bronze at first, becoming a darker green: the perfect foil to spherical flower heads that flounce on to the scene in high summer. The petals are white, distinctively fringed with pink, and are carried erect on strong stems that lend themselves to cutting.

As the flower ages, it becomes creamier in color before antiquing to lime, tarnished copper and pistachio in an exit that is just as fabulous as its entrance.

With frilled and picotee-edged hydrangeas fantastically popular at present, there is a whole range of dangerous and dissolute-sounding blooms to create a garden-party vibe, should one so desire. These include outrageous, fuchsia-toned French Cancan and seductive Love You Kiss (see overleaf).

Hydrangea macrophylla 'Cocktail'
Height 4 ft–5 ft 3 in
Spread Up to 5 ft
Flower size Medium
Habit Makes a substantial bush
Color Fresh white flowers with fringed cerise margins
Preferred situation Partial or dappled shade in a spot that does not get too cold
Hardiness USDA Zones 6–7
In the garden An interesting new addition
As a cut flower Pretty in an arrangement with dahlias and old roses, perhaps with cosmos, gypsophila or *Panicum* grasses to soften the effect

ABOVE: FRENCH CANCAN RIGHT: LOVE YOU KISS

Curly Sparkle Purple

If your tastes tend to the weird, wonderful or just plain bizarre, you can't get much more curious than the Curly® Series of hydrangeas.

Available as Curly Sparkle Blue, Curly Sparkle Pink and Curly Sparkle Purple, the buds open apple green and tightly frilly before expanding and taking on their mature colors. And even when they are fully developed, they remain fanciful and inconsistent, the sepals veined with a paler shade, while contrasting hues bleed from the center of each floret. As it ages, the whole flower becomes darker and more intense.

With this fluid approach to color and decisively wavy edges, this mophead looks less like a classic, tidy, well-behaved hydrangea and somehow more organic. Like a plant that has lost its way, whose real calling is to be under the ocean waves, attached to a rock alongside the other sessile creatures – anemones, barnacles and polyps – basking in diffused sunlight and being nibbled at by fish.

In the garden, show-offs like this should be given pride of place, but don't expect consistency; like all macrophyllas, the underlying soil type will influence the color so it may come up bluer or pinker – it's just the way it rolls.

....

Hydrangea macrophylla 'Curly Sparkle Purple'
Height 40 in
Spread 40 in
Flower size Medium
Habit Domed, deciduous shrub
Color Broadly purple; Curly Sparkle Blue and Curly Sparkle Pink are also available
Preferred situation Like all macrophyllas, it likes some shade and not too much cold
Hardiness USDA Zones 6–7
In the garden An attention-seeking show-off, probably best in a container where it can wow its audience
As a cut flower Talk of the town!

Invincibelle Spirit II

There is something almost cartoonish about Invincibelle Spirit II; the shrub is not overly large, but it is smothered in flowers – large, rosy globes, all bouncing, bodacious and thoroughly magnificent.

Invincibelle Spirit II comes from the same stable as the original pink form of Annabelle and it has many of the same endearing characteristics. There are the strong stems that seem to have no problem in holding the floral beach balls aloft; it also repeat-blooms on fresh shoots to give a long season of interest. Additionally, it thrives in a wide variety of gardening climates – happy enough in chilly regions, yet also tolerant of more tropical climes.

The flowers are the sort of pink that means business – neither faded nor pale, and far too robust to be described as sugary. The color deepens with age, taking on green tones with a hint of metallic teal.

It is also a flower with a conscience: in the USA, one dollar from each plant sold is donated to The Breast Cancer Research Foundation®.

....

Hydrangea arborescens 'Invincibelle Spirit II'®
Height 4 ft
Spread 4 ft
Flower size Very large
Habit A deciduous, medium-sized bush
Color Delightful ice-cream pink
Preferred situation Partial sun to full sun
Hardiness USDA Zone 3
In the garden Striking and easy-to-care-for shrub
As a cut flower Arrange when fresh if you have a container large enough, or cut and dry the blooms

Endless Summer The Original

When the first Endless Summer® hydrangea appeared, it was met with much excitement. Here, finally, was a plant that would rebloom, producing wave after wave of flowers on new wood all summer. This meant that it could be grown successfully in colder situations than could other *macrophylla* varieties.

Since then, the breeders have developed a collection of hydrangeas under the Endless Summer banner, including white Blushing Bride and lacecap Twist-n-Shout (see page 64). The Original is a handsome mophead that is classically blue, but will turn a pleasing pink in gardens with alkaline soil.

Endless Summer hydrangeas are easy to grow as long as they have good soil and plenty of moisture, and their resistance to powdery mildew is also an asset. They are best in partial shade as they will scorch if they receive too much sun.

Hydrangea macrophylla Endless Summer® 'The Original'
Height 4 ft
Spread 5 ft 3 in
Flower size Large
Habit Domed, spreading deciduous shrub
Color Blue or pink
Preferred situation Like all macrophyllas it likes some shade and not too much cold
Hardiness USDA Zones 4–7
In the garden A reliable rebloomer that give you lots of bang for your buck
As a cut flower Big globes of flowers that really make a statement

Rembrandt Vibrant Verde

Gardeners have an insatiable lust for the new and glamorous. Fresh color-breaks, different forms, twists in the tale – when you are tired of gardening you are tired of life.

Yet, with hydrangeas it might be easy to think that one has achieved an elegant sufficiency. A pleasing collection of species and forms that fulfills your every horticultural need, without tipping over into a state of full-blown addiction.

And then along comes the Rembrandt® Series. They are clearly named for the notable artist, or more specifically to evoke the myriad hues that might spring from his palette, and they drive the point home by morphing gaily throughout the season.

Thus, Rembrandt Vibrant Verde is a compact and versatile little shrub, pumping out small, arsenic-green flowers with enthusiasm until the whole plant is smothered. The center of each floret starts off white before turning pink. This rather idiosyncratic look lasts just moments, however. The pink stain wells up dark and dominant to flood the flower and finish the year ruddy and ensanguined.

While this shrub will perform well in the garden border, its attractions are such that it is better used as a container subject on a deck, balcony or patio, where it will thrive in a shady spot.

....

Hydrangea macrophylla Rembrandt® 'Vibrant Verde'
Height Up to 40 in
Spread Up to 40 in
Flower size Small
Habit Compact and bushy shrub
Color Snake-green flowers with a white eye
Preferred situation Partial or dappled shade; avoid severe cold
Hardiness USDA Zones 6–7
In the garden Great in containers and tumbling on to the patio in pride of place
As a cut flower Eye-catching, especially when teamed with purples and oranges; dries well

Brilliant
and Bold

Hamburg

With a plethora of ever more tantalizing, newly minted hydrangea varieties laid out before us on an annual basis, it is worth remembering that a good plant tends to stick around.

The old familiar names are usually popular for a reason and so it is with Hamburg. Introduced by a Herr Schadendorf in Germany in 1931, this is a plant that has had nearly a century to prove itself – and it may just be the ultimate mophead hydrangea.

The heads of densely clustered florets are enormous; big bold globes in hues that range from mauve-rose and lilac on alkaline soils to the clearest of blues on an acid soil – the lower the pH the more intense the color.

Like all *macrophylla* types, it likes a warmish site and plenty of moisture underfoot, although, anecdotally, it may be more tolerant of cold than some. Deadhead and prune out crowded stems in late spring. To beef up the size of the blooms, prune out some (or lots) of the flowering buds and the plant will invest all its energy in those remaining, which will result in the most magnificent flowers imaginable.

If Hamburg has a fault, it may be that it doesn't present as well potted up for sale as some of the modern varieties. So, where the market is dictated by fashion rather than gardening, it can be harder to find, but seek it out and plant it in the ground rather than in a container, and you will be glad that you did.

....

Hydrangea macrophylla 'Hamburg' syn. *H. hortensia* 'Hamburg'
Height Up to 6 ft 6 in
Spread Up to 8 ft
Flower size Large
Habit Medium to large shrub
Color The flowers are rose pink to clearest blue, depending on the soil pH
Preferred situation Partial shade or full sun in a sheltered spot
Hardiness USDA Zone 6–7
In the garden Lovely in a woodland garden or mixed border but less successful in containers
As a cut flower Works well in substantial fresh arrangements, particularly in autumn, when it turns wine-red; it also makes an excellent dried flower

Alpenglühen syn. Alpen Glow, Glowing Embers

Bred in Germany in 1950 by a gentleman by the name of Herr Brügger, Alpenglühen remains one of the best red-colored hydrangeas available.

Robust in growth but not too large, it is tolerant of cold or sunny sites that would have other hydrangeas of this species running for cover, and it will thrive in a coastal spot, too. The large mophead flowers are produced over a long period and are one of the deepest in color of any hydrangea of its type.

The green buds rapidly take on pink tones as they expand, the color bleeding from the petal tips towards the center and deepening to rich tones of cerise and carmine. In age, the flowers finally become red-bronze and then finally bronze-purple, hues echoed by the leaves, which flame to maroon before they fall.

Unlike many *macrophylla* cultivars, Alpenglühen does its best to remain stable regardless of soil pH – it is a red that really *wants* to stay red and it will only succumb to the most acid of soils. Even then it changes color only grudgingly, to become damson or mauve.

Alpenglühen translates to English as 'Alpine Glow' but it is also known as Glowing Embers and, somewhat unimaginatively, Forever Pink.

Plant Alpenglühen where its strong color and globular form can be appreciated, as a specimen or in a perennial or hot border, where it will contribute structure and substance.

....

Hydrangea macrophylla 'Alpenglühen' syn. *H. m.* 'Alpen Glow', *H. m.* 'Glowing Embers'
Height Up to 4 ft
Spread Up to 4 ft
Flower size Large, up to 8 in across
Habit Small to medium shrub
Color Deep pink-red
Preferred situation Partial shade or full sun
Hardiness USDA Zones 6–7
In the garden Flowers over a long period; grow in pots or use as a specimen shrub
As a cut flower Lasts well in a vase; try combining it with tawny grasses or dark plum flowers and foliage

Wim's Red syn. Fire and Ice

This delectable new hydrangea is rapidly becoming a garden must-have and with good reason. Firstly, the flowers are precocious, arriving earlier than most *paniculata* varieties, and sometimes as soon as late spring in an established plant.

And, secondly, the coloring is captivating – the substantial-yet-airy plumes of flowers start off a creamy white, set off beautifully by strong maroon stems that are more than equal to the task of holding the flowers erect. As the season progresses (see pages 174–175), the blooms take on a rosy flush that deepens and intensifies to an inky magenta, then rich maroon, before finally antiquing to pewter. To top off this list of assets, the flowers are deliciously honey-scented and the leaves yellow attractively in autumn.

A boon to substantial arrangements, the flowers also look good dried. In the garden, Wim's Red makes an eye-catching specimen shrub and it can also excel in a mixed border, where it mingles well and provides a long season of interest. For the best colors, make sure that it gets plenty of light, but choose a spot that doesn't get too hot; it will also resent its feet drying out.

....

Hydrangea paniculata 'Wim's Red'® syn. *H. p.* 'Fire and Ice'
Height Up to 6 ft 6 in
Spread Up to 6 ft 6 in
Flower size Large
Habit Medium to large shrub
Color Fresh cream flowers gradually assume a glorious fiery glow
Preferred situation Happy in partial shade but the colors will be more intense in sun
Hardiness USDA Zones 3
In the garden Striking wherever it is sited and provides interest over a long season
As a cut flower In a large vase, Wim's Red will look great early in the season combined with fresh greens such as *Mollucella,* green amaranth and nicotiana, and white cosmos; later in the season, beef up the display with plum colors or maybe even go wild with a hit of burnt orange

Miss Saori

There is something unutterably enchanting about bicolored hydrangeas, and the lavish and richly ruffled mopheads of Miss Saori are enough to captivate the most indifferent observer and would melt the stony heart of the most committed hydrangea-hater, should such an individual conceivably exist.

As a garden plant, she is magnificent, a pint-sized floral explosion that performs from late spring until well into the autumn, when the already plum-soaked leaves join in the party with a display of ruby and bronze. The tiered flowers are made up of pointed sepals, white in the center and with the edges dipped in raspberry pink, a color that remains reasonably consistent regardless of soil pH.

This unique plant was bred in Kyoto by hydrangea fanatic and rock guitarist Mr Ryoji Irie, who has been experimenting and tinkering with plants for years. It is a hybrid of undisclosed but clearly promising parentage and it is named for his wife.

Part of the You & Me® Series (see page 180), Miss Saori was winner of the RHS Chelsea Flower Show Plant of the Year in 2014.

....

Hydrangea 'Miss Saori'
Height 40 in
Spread 40 in
Flower size Medium
Habit Compact little bush
Color Striking white-and-pink flowers
Preferred situation Partial or dappled shade, or sun, as long as there is plenty of moisture underfoot
Hardiness USDA Zones 6–7
In the garden A vivid and adaptable specimen shrub
As a cut flower Excellent; the rich berry colors work well in informal arrangements or in a relaxed posy with Japanese anemones, sweet williams, clary sage and scabious

Merveille Sanguine syn. Brunette, Folis Purpureus

One of the richest, darkest red hydrangeas going, Merveille Sanguine was discovered as a sport, or mutation, of the old, pink hydrangea variety Merveille, and was introduced by Henri Cayeux in 1936.

And it is a plant that is fabulously striking. The large, regular mopheads are made up of deep-blackcurrant flowers, each with a central eye of vivid violet. The flowers are carried over glossy foliage that has been washed with bitter chocolate, the leaves starting their lives a deep greenish-bronze color and then retaining a purple cast as they mature, before finally capitulating to inevitability and committing to a good autumn hue. The overarching sense is of a plant that is a controlled explosion of crimson, cocoa and wine intensifies and deepens throughout summer until it reaches its flaming autumnal conclusion.

Merveille Sanguine is an excellent cut flower: use in an autumn arrangement with rosehips, hawthorn berries, scabious, maple leaves, chrysanthemums and amaranth.

Leave the flower heads on over winter to protect the underlying buds and plant it in a warm microclimate if your garden tends to be cold.

Known also as *Hydrangea macrophylla* 'Brunette' or 'Folis Purpureus', the name Merveille Sanguine translates literally as 'Marvellous Blood'. But it rolls better off the English-speaking tongue as 'Bloody Marvel', which many connoisseurs of the genus would consider both nominative determinism and accurate naming firmly out in force.

....

Hydrangea macrophylla 'Merveille Sanguine' syn. *H. m.* 'Brunette', *H. m.* 'Folis Purpureus'
Height Up to 5 ft
Spread Up to 5 ft
Flower size Medium
Habit A rounded, medium-sized shrub
Color Glowing crimson-wine flowers that deepen with age
Preferred situation Partial or dappled shade in a spot that does not get too cold
Hardiness USDA Zones 6–7
In the garden A vivid specimen shrub
As a cut flower Excellent in seasonal arrangements, the flowers are also good dried

Together

When florists go to bed at night, they offer up a little prayer. They pray that people will love their bouquets, obviously, but more than that, they dream of flowers that are sumptuous and versatile. That have something 'other' about them. Flowers that, these days, would be said to have the X factor and, 40 or 50 years ago, would have been said to have 'it' – that indefinable quality of desirability that brings the crowds flocking and keeps the money rolling in.

The gods of floristry seem to have been listening and, up in their celestial hybridization unit, they came up with Together – a hydrangea that must surely have been crafted by deities and delivered by angels.

The mopheads are huge and lush, made up of exquisite double flowers that open light green, become blue on acid soil and eventually assume a hint of violet as they fill out and mature. The evolving tints are subtle and forgiving, mixing well with pastel colors and white, both in bouquets and in the garden, with the blue a gently luminous foil to the surrounding brilliance.

If you don't want to cut the flowers, remember that *macrophylla* hydrangeas like a good level of soil moisture when grown outdoors, and prefer some shade as they may scorch in hot sun. Acid soil will produce the truest blues.

Together is one of the You & Me® Series that also includes Love, Passion, Forever and Expression (see overleaf) – although they do look rather different. These, plus Inspire (see page 192), were bred by Japanese czar of guitar and gardening, Mr Ryoji Irie, and one thing is certain – they will rock you.

....

Hydrangea macrophylla 'Together'
Height 40 in
Spread 40 in
Flower size Large
Habit Small shrub
Color Shades of blue or pink, aging to crimson-burgundy
Preferred situation Like all macrophyllas, it likes some shade and not too much cold
Hardiness USDA Zones 6–7
In the garden Eye-catching in containers
As a cut flower Big globes of flowers have real presence and also work well dried

183   LEFT: TOGETHER   ABOVE: EXPRESSION

Enziandom syn. Gentian Dome

In the mind's eye, gentians are synonymous with alpine pastures, melting snows and goatherds each with a feather in their cap – and they are also renowned for being a truly intense and glorious shade of blue.

But while gentians themselves can be tricky customers in cultivation, not so *Hydrangea macrophylla* 'Enziandom'. And, given a decent soil and some afternoon shade, it will flower its socks off.

The clear blue orbs are bold, solid and heavy, but the stems are stout enough to be up to the job of supporting them, forming a static cascade of color. The sepals of each floret curve smoothly to a point and morph from the cerulean hues of summer to a more steely tone as they reach their antique phase.

The only issue with this particular variety is that, like others of its species, it becomes muddied with pink on soils with a higher pH. But on acid soil, it is a thing of unsurpassed beauty.

....

Hydrangea macrophylla 'Enziandom' syn. *H. m.* 'Gentian Dome'
Height Up to 4 ft
Spread Up to 4 ft
Flower size Medium to large
Habit A rounded, medium-sized shrub
Color In good acid soil, a vivid bright blue
Preferred situation Partial or dappled shade in a spot that does not get too cold
Hardiness USDA Zones 6–7
In the garden An eye-catching feature plant
As a cut flower Bold and dramatic; dries well to a pleasing metallic blue

Harlequin

Adding an element of pantomime to the garden, Harlequin is nothing if not extrovert.

The flowers are eye-catching and rather jazzy. Each sepal is rimmed with white around a central, colored rhombus, creating a diamond-patterned effect. But while some people find this sensational, others are less enamored, considering it to lack subtlety. Indeed, it could be accused of brashness.

But while it can be tricky to blend in with other planting, its statement value is beyond question. When the heart of the petal is a good, clear shade of pink or blue, it has a unique kind of elegance, almost confounding expectation. And, if necessary, its more outrageous tendencies can most likely be toned down by surrounding it with rich shades of green.

With a rather delicate constitution, it can struggle to thrive and perform. Indeed, it may be something of a labor of love to cosset and tend it through its consumptive periods.

But if you have a shady spot on a sheltered patio, and if what you want is a plant with glamour, pizzazz and personality, and that is not afraid to show its true colors, then Harlequin is the hydrangea for you.

Hydrangea macrophylla 'Harlequin'
Height Up to 4 ft
Spread Up to 4 ft
Flower size Medium
Habit A mounded shrub
Color Variegated, pink to mauve with white edges
Preferred situation Partial or dappled shade; avoid frost pockets
Hardiness USDA Zones 6–7
In the garden A striking plant for a border or container
As a cut flower Eye-catching; the flowers dry well

Munchkin

One of the nice things about the *Hydrangea quercifolia* cultivars is their leaves. While other hydrangeas have foliage that could politely be described as "pleasant" and which is fairly ordinary in appearance, with quercifolias, the leaves are just as much of an asset as the flowers – nicely lobed in shape and prone to flare a brilliant and spectacular scarlet, crimson and mahogany in autumn.

With a long and varied season of interest, Munchkin is a good sort of shrub to choose for a small garden. It is compact, unfussy and nicely proportioned, bearing loose, ivory flowers that take on a pink color as they mature.

Rather than using it as a single specimen, try surrounding it with grasses such as *Carex* or *Pennisetum* species to soften the look and give it some more context, and perhaps a more designed feel. As it is deciduous, you could also underplant with choice bulbs such as snowdrops or cyclamen to extend the season of interest.

....

Hydrangea quercifolia 'Munchkin'
Height Up to 36 in
Spread Up to 4 ft
Flower size Medium
Habit Spreading, domed shrub
Color White panicles of flowers gradually develop a hint of pink
Preferred situation Partial shade or full sun
Hardiness USDA Zone 5
In the garden Provides flower and foliage interest over a long period and while it can be grown as a specimen shrub or in a mixed border, it is also compact enough for containers
As a cut flower Dries well and retains pink hues when cut in early autumn

Inspire

Modern, gorgeous and completely captivating, Inspire is one of that stampede of new hydrangeas that are doing so much to ensnare hearts and rebrand the entire genus *Hydrangea* as a must-have.

The full, mophead flowers are a rich and glorious pink color, and the elongated sepals are stacked on top of each other to create textured plumes. Inspire has a flower that speaks of sensuality and abandon: reminiscent of the feather boa joyously sported by a drag queen on a fabulous night out, or rose petals cascading fragrantly and sumptuously in the boudoir.

In the garden, given a consistent amount of summer moisture, Inspire is a relatively low-maintenance plant. Acid or alkaline soil will alter the color, making the flowers more violet or more pink respectively. It also produces new blooms all summer for a long and glamorous display. While happy in the border, this plant can look stunning replicated in matching pots for a contemporary take on a classic look.

Hydrangea macrophylla 'Inspire'
Height 40 in
Spread 40 in
Flower size Large
Habit A compact, domed plant
Color Pale pink flowers deepen with age
Preferred situation Partial or dappled shade in a spot that does not get too cold
Hardiness USDA Zones 6–7
In the garden An eye-catching addition
As a cut flower Cut and present to a loved one as a declaration of passion, or use dried in wreaths and garlands

Doppio Rosa

This little plant is quite simply gorgeous. Rose-pink lacecap flowers are borne above a mound of greenery and gradually expand like stars appearing in a darkening sky.

Slowly, slowly, the little florets made up of layers of neat and pointed sepals start to fill up the center of the corymb, bulking out the already not-insubstantial flowers and giving the plant more presence as the season wears on.

Versatile and easy to grow, Doppio Rosa can flower from late spring until autumn when it gets going, and it is compact enough to make a nice feature in a large pot, perhaps near an entrance or used in pairs, either side of a flight of steps.

If you like the 'little bunch of fireworks' aesthetic but pink is not your thing, take a look at Doppio Rosa's sister shrub, Doppio Bianco, now known as Wedding Gown (see page 58), which does the same thing but in white.

Hydrangea macrophylla 'Doppio Rosa'
Height Up to 40 in
Spread Up to 40 in
Flower size Medium
Habit A compact and tidy shrub
Color Pink
Preferred situation Partial or dappled shade in a spot that does not get too cold
Hardiness USDA Zones 6–7
In the garden A really dramatic small specimen and good in containers, too
As a cut flower Dainty and not too overblown

Nikko Blue

If you have acid soil, Nikko Blue is one of the best hydrangeas about, and even on lime soil and with the flowers in a rosier incarnation, it is pleasant enough.

A well-known and established cultivar, it has smooth heads of overlapping florets rather resembling a ladies' swimming cap from the 1950s. It is also remontant, which means that it flowers over a long period and is more frost-resistant than single-blooming cultivars.

Plant Nikko Blue where it can be most appreciated. It works well as an informal hedge along the edge of a lawn, or try siting it under mature trees with companions such as ferns and hostas, or with other plants that have a light or luminous quality such as irises and astilbes.

....

Hydrangea macrophylla 'Nikko Blue'
Height Up to 5 ft
Spread Up to 5 ft
Flower size Medium
Habit A rounded, medium-sized shrub
Color In an acid soil, beautiful pale blue
Preferred situation Partial or dappled shade in a spot that does not get too cold
Hardiness USDA Zones 6–7
In the garden A trouble-free border plant
As a cut flower Elegant and refined

Hopcorn

Some novelty plants are bred by assiduous and deliberate crossing of known parents but others arise spontaneously, due to a sport or mutation that is pounced upon and perpetuated. Such is the case with Hopcorn – and the story is rather sweet.

Hopcorn was discovered and bred by father-and-son team, Koos and Wilko Hofstede, in the Netherlands. The story goes that Wilko was sorting out his plants one autumn when he noticed that some of the florets had an odd, rather cupped and chubby appearance, resembling those of a lilac.

Intrigued, the pair bred and selected from this plant, over the years refining the genes until the entire flower head was made up of the unusual florets. They were then able to propagate from the plant and think of a name. The crumpled, deeply cupped petals reminded them of popcorn and their name is Hofstede, so they combined the two to get Hopcorn – 'Hofstede's Popcorn', to go with the Hovaria® Series of special hydrangea cultivars – a conflation of 'Hofstede's Variation'.

Now commercially available, the mophead flowers are similar to those of Ayesha (see page 134), but the plant is rather smaller. On acid soil it is a glorious blue color and it goes through shades of purple to a dark pink on calcareous soil. Sometimes marketed as 'Hopcorn Blue', the "blue" epithet is to indicate that it has been grown in acid soil and with aluminium treatment, so the color is assured even if the flower is not yet out.

Tolerant of heat and cold by *macrophylla* standards, it still does best planted in decent soil and with a bit of shade. Indoors, it likes bright but diffused light and a cool location.

....

Hydrangea macrophylla Hovaria® 'Hopcorn'
Height Up to 4 ft
Spread Up to 4 ft
Flower size Medium
Habit A compact, medium-sized shrub
Color Flowers that are a lovely blue on acid soil and coral pink on lime
Preferred situation Partial or dappled shade in a spot that does not get too cold
Hardiness USDA Zones 6–7
In the garden Something a little bit different
As a cut flower Dense flower heads combine well with other hydrangeas or loose, vintage-style roses

HYDRANGEA GROWING AND CARE

THE HUGE NUMBER OF HYDRANGEAS SOLD AS HOUSE PLANTS AND GARDEN SHRUBS IS A TESTAMENT NOT JUST TO THEIR POPULARITY, BUT TO THEIR EASE OF CULTIVATION. YET, WHILE THEY MAY BE LOW-MAINTENANCE, THEY DO HAVE NEEDS AND PREFERENCES. SELECT AN IDEAL SPOT, AND DEPLOY A LITTLE TARGETED CARE IN THE FORM OF PRUNING AND FEEDING, AND A HUMBLE SHRUB CAN BE ELEVATED FROM THE ORDINARY TO THE TRULY MAGNIFICENT.

Cultivation

By and large, hydrangeas are tolerant plants, but the genus is extensive and fairly widespread. Knowing where key garden species originally came from and aspiring to replicate those conditions is, therefore, the fastest route to glory.

And to do your best by your plants, you also need to become familiar with the conditions in your own garden. Give yourself time to get to know your climate and soil; think about where the sun falls and see where the ground has a tendency to hang wet or dry out.

It is possible to make general statements about growing hydrangeas. For example, they tend to like a good, consistent supply of water, particularly in hotter or dryer areas. They will often grow quite happily in part shade – or even full shade, in some cases – but, as a rule of thumb, the more sun they get, the more moisture around the roots they will need to do well.

Each hydrangea species is a product of its ancestral home. The coastal, frost-free *macrophylla* types are more tender, while the *serrata* species from higher up the mountains tend to be a bit more cold-tolerant. Inland species such as *Hydrangea quercifolia*, *H arborescens* and *H. paniculata* are increasingly hardy, culminating in an ability to survive serious cold.

To cultivate hydrangeas really successfully is a bit like becoming a plant-whisperer. To grow as one with the specimen before you. To know the soil beneath your feet and the shade cast from above. To understand the need for both drainage and plenty of moisture; and to appreciate the scorching or color-enhancing effects of the sun.

But the hydrangea is a kind and tolerant teacher. It will allow you to make mistakes on the road to understanding and enlightenment. It is a plant that will put up with relocation and largely forgive transgressions as regards sun, water and frost. You may lose a few flowers, but the chances are that it will bounce back the following season for another attempt at splendor.

In short, they are easy and rewarding plants to cultivate so, with your growing conditions in mind, go forth and seek a hydrangea to suit. With so many assets and so few drawbacks, it should be a sure-fire winner.

Selecting varieties

Hydrangeas vary in their tenderness and tolerance of sun, salt and moisture levels, so it is simplest to select one to suit your region. But rules are, to some extent, meant to be broken and there are a few natty tricks you can use to cheat the conditions if you wish.

A key element in plant choice is its ultimate size. Many newer varieties are bred to be extremely compact, perfect for containers and courtyards or for using as a house plant. Others make excellent specimens or landscape shrubs, while still others become whopping climbers or small trees. As with location and climate, it pays to do a bit of research as to the scale of the plant you have your eye on, to make sure that it will fit into your long-term plans.

Below I outline some of the main species and their requirements and characteristics. You will find more details on size and color of a number of key cultivars in the profile chapters.

Hydrangea macrophylla

Sometimes known as 'bigleaf hydrangeas' with flowers that are mophead or lacecap, according to cultivar, *Hydrangea macrophylla* is native to coastal regions of Asia. While this can bring its own challenges, such as exposure and salt-laden winds, the relatively stable temperature of the sea acts as a buffer – if temperatures drop, it acts a bit like a radiator and, if they rise, it provides air cooling. The adjacent land is thus less prone to frost than further inland and *H. macrophylla* had little need to develop a tolerance to cold.

H. macrophylla also does best in semi-shade, perhaps with sun for half the day, or dappled sunlight. It will also grow well under deciduous trees with no direct sunlight at all – although in total shade it will be leggier and skinnier than usual, and will produce fewer flowers.

Charmingly versatile, it will also grow in full sun. But only, absolutely only, if you can give it plenty of water at its roots – it will even take a boggy site if it has to. The leaves of *H. macrophylla* are large, as are the often-lavish flowers, and this sumptuousness takes a lot of moisture to sustain.

This is a very useful garden plant indeed. It is tolerant of wind and salt, and full shade to bright sun, and although it prefers a wet climate, it compensates for this foible by also putting up with wet soil. But it simply won't take cold.

Gardeners are, however, optimists so although *H. macrophylla* is 'officially' hardy to USDA Zone 7, or 6 at best (see the table on page 209), in places where the winter temperature dips below about -4°F, around the middle of USDA Zone 6, people have come up with all sorts of ways to try and coax their hydrangeas through.

A warm microclimate, close to a wall or near larger, sheltering plants can help. It is really important to avoid frost pockets and pay attention to what is known as "air drainage," where dense, cold air effectively pours downhill to pool at the bottom – meaning that an uphill site can be markedly warmer. Delaying pruning until spring, meanwhile, means that the old flowers can, to some extent, protect the buds of the new ones, trapping a layer of air within the plant and sheltering the buds from the top.

In colder areas still, further measures are required. In winter, the dormant plants don't need much light, so tucking your hydrangea up in an old blanket, or under several layers of sacking, may get it through. Alternatively, you can build a cage and fill it with an insulating layer of leaves and straw, and then remove it in spring – covering with a quilt if spring frost is forecast.

But when growing *H. macrophylla* in a cold climate, surely the most foolproof way must be to keep the plant in a container and move it into a frost-free space when the leaves drop.

Fortunately, the clever boffins in hydrangea development have been working on a solution. Hard frost kills *macrophylla* buds and stems, and, because it is these parts that produce the flowers, too much frost means no flowers. (Really extreme frost will kill the roots too, but these are protected significantly by the soil, and defenses can be beefed up by adding a thick layer of mulch.) A few varieties of *H. macrophylla* bloom more than once, for example Générale Vicomtesse de Vibraye and Nikko Blue. These are known as remontant types and some are more remontant than others. The search has been on to discover, breed and select more of these reblooming hydrangeas and newer varieties include Penny Mac, the expanding Endless Summer® Series and Runaway Bride® 'Snow White', which has opened up the opportunities for cold-climate hydrangea-growing considerably.

Hydrangea serrata

A near neighbor and relative of *H. macrophylla*, *H. serrata* hails from the wooded mountains of Japan and Korea, where it is sometimes called "tree of heaven." In the uplands, well away from the sea, temperatures often dip below freezing so, for garden purposes, it is noticeably hardier than its cousin, and can be grown in regions that reach -10°F, or USDA Zone 6.

While the vulnerability of *H. macrophylla* comes from the length of its growing season, starting too early and going on too late, *H. serrata* has a much shorter growth period so it is hardier in the face of cold – although it is no fan of blazing sun.

Like *H. macrophylla*, *H. serrata* likes partial or dappled shade so it can be used in a similar way in the garden. It is less tolerant of exposed, windy or very warm locations and it doesn't thrive in wet soils, which makes it less suited to coastal and boggy sites. On the other hand, the plants tend to be smaller, at around 40 inches tall, so are suitable for compact modern gardens.

The species has lacecap flowers and serrated leaves – hence the name – and does well under trees. A number of cultivars, specifically Grayswood, Preziosa and Glyn Church will go through several color changes throughout the season – but since they are not susceptible to pH, these are consistent in their inconsistency. The white cultivars will remain white regardless of soil pH, but the other pink and blue cultivars are moderately susceptible, so situations arise where, for example, Bluebird, grown on alkaline soil, will produce flowers that are noticeably pink.

Hydrangea quercifolia

Given a choice spot in a warm and sunny garden, *H. quercifolia* is a plant to be reckoned with. Native to south-east America, it has panicles of white flowers that form erect spires or cascading avalanches. These are set off by the lobed foliage, which is an attractive green in summer but really comes into its own in autumn, when it takes on tones of copper, crimson and rich, inky purple.

REMONTANT HYDRANGEAS

Typically, *Hydrangea macrophylla* flowers on old wood. This means the stem that grew and developed buds last summer and autumn, and then stopped for winter. In spring the buds expand and start to grow again – this new growth being what actually carries the flower.

As a consequence, the pruning advice has always been to cut back to a nice fat bud near the top of the stem, in mid-spring.

The problem is that, if a classic *macrophylla* is pruned back hard, hit by frost that kills the stems to the ground or eaten by grazing herbivores such as sheep or deer, the plant has to start again, producing a stem, then a bud, in which develops a flower that still has ambitions to bloom that year. And in a cold climate with a short growing season, many varieties simply don't have time to complete the process – ergo, no flowers.

Where remontant or reblooming types differ is that they have the firepower and precocity to produce flowers fast, throwing out shoots that will flower at the top on fresh growth, without a winter break and whether or not they are cut to the ground. This decorative and useful capacity is particularly helpful when growing hydrangeas in a cold climate and can be found in a number of *H. macrophylla* and *H. serrata* cultivars.

The leaves resemble those of an oak – particularly the American red oak – so the plant is commonly known as the oak-leaf hydrangea, and it really does thrive in a sunny spot, although a hint of shade is a good idea in places where the heat gets extreme.

By the standards of its kind, the plant is moderately hardy. It will survive winter lows of approximately -22°F, or USDA Zone 5, as long as summers are warm. It responds well to being grown in local microclimates that stop the temperature dipping too low, but it doesn't really perform to its best in shade and it also dislikes very dry conditions. It also can be a bit marginal in places where the summers are prone to be cool.

Hydrangea arborescens

The first hydrangea to be introduced into cultivation, the garden form of *H. arborescens*, at only about 40 inches tall tends to be distinctly less tree-like than its name might suggest, but it is graceful, nevertheless.

Native to North America, its wild form can be found over a wide swathe of eastern America, running down the Appalachian Mountains from New York to Florida. It is, therefore, adapted to a somewhat mountainous life, where things are either cool and damp or bright, warm and dry, but never hot and wet.

As a result, unlike *H. macrophylla*, it doesn't really like to sit in a damp spot in a warm garden and extra attention should be paid to drainage. On the plus side, it is fearless in the

face of severely cold weather and it will tolerate drought very much better than other species. With changing weather patterns, *H. arborescens*, together with *H. paniculata* are increasingly preferable choices, over the more vulnerable *H. macrophylla*, in challenging (or extreme) garden design conditions.

Hydrangea paniculata

Originating in Japan and south-east China, *H. paniculata* is another hardy species, and it is as handsome as it is trouble-free.

The pointed flowers are often spectacular and can be dense or airy, depending on the cultivar. It tends to form a large shrub or even a small tree, potentially reaching 20 feet in a favorable site, although usually very much smaller. It prefers to be grown in full sun and it is unfussy about soil; it naturally prefers rich pickings and good drainage with plenty of moisture, but if what is on offer is heavy clay or unpromising pebbles, it will make the best of the situation as long as it is not waterlogged, and it will also put up with drought.

Like *H. arborescens*, it is also very cold-tolerant, down to USDA Zone 3, which represents a chilly -40°F or so, and, with its large flowers and dramatic good looks, it is a very good alternative to *H. macrophylla*.

If it has a flaw, it is that the branches tend to be brittle, so it is best planted in a sheltered situation as, in a stiff breeze, they may not be sufficient to support the flowers.

HARDINESS

There is a range of charts and tables to indicate what level of chill a plant will tolerate. Two useful systems are produced by the United States Department of Agriculture (USDA) and the Royal Horticultural Society (RHS), which categorize plants from tropical to extremely hardy.

USDA
Zone 3: -40°F to -30°F
Zone 4: -30°F to -20°F
Zone 5: -20°F to -10°F
Zone 6: -10°F to 0°F
Zone 7: 0°F to 10°F
Zone 8: -12°C to -7°C (10°F to 20°F)
Zone 9: -7°C to -1°C (19°F to 30°F)

RHS
H3 to 23°F half-hardy
H4 to 14°F hardy in an average winter
H5 to 5°F hardy in a cold winter
H6 to -20°C (-4°F) hardy in a very cold winter
H7 colder than -20°C (-4°F) very hardy

Because these systems do not equate exactly, the specified hardiness in the plant profiles on pages 38–199 are an approximation; local variations in climate will also play a part.

Hydrangea petiolaris and other climbing hydrangeas

Whereas most hydrangeas are shrubs or small trees, *H. anomala* subsp. *petiolaris* is a magnificent and hardy climber, which can take temperatures as low as the vast majority of climates will throw at it and still cover itself in airy, white, lacecap flowers.

In the wild, it is found in Korea scrambling up whatever vertical surfaces it can find – cliff faces, rocks and trees – to which it clings with short adhesive roots, a bit like ivy. It enjoys good living as much as the next shrub, but it will tolerate poor soil, dry conditions and shade, which makes it an excellent plant for unprepossessing urban situations.

H. petiolaris can get very large if given something substantial to climb up, but in general it will grow to the extent of the available support and then stop.

There are a number of other climbing species, too and, of particular note, are *H. seemannii* (see page 72) from Mexico and *H. serratifolia* from Chile. Both of these are evergreen and rather frost-tender, although they can thrive if the microclimate is suitable. You may also come across the Japanese hydrangea vine, S*chizophragma hydrangeoides*, but although this is related and superficially similar, it is not a true hydrangea.

Other hydrangeas

As discussed earlier, the genus *Hydrangea* is a large one and has not been fully explored. A number of good species are relatively uncommon in cultivation, and others have not reached gardens at all, although some breeders are using wild forms in an attempt to introduce exciting new variables, such as reblooming, hardiness or richer colors.

If exploring a superior nursery or browsing online, you may come across the tall and elegant Himalayan hydrangea, *H. heteromalla*. Size-wise this can be a bit of a beast, with large, hairy leaves and compound flowers made up of individual flower heads, so it is really best in large gardens. It also varies in hardiness, depending on which form you find.

H. involucrata is also a fascinating variation, with large felted leaves and flower buds that present themselves as big, round, mossy balls, rather like those of a peony.

Another more unusual variety is the Chinese species, *H. aspera* (see page 116), which has lately been extended to include a number of plants that were previously designated as species in their own right. So beautiful *H. villosa* is now termed *H. aspera* (Villosa Group), to cover all the variations. Likewise, *H. sargentiana*, a choice although sometimes challenging variety with velvety leaves and large flowers, is now known as *H. aspera* subsp. *sargentiana*.

In general, try to pick a named cultivar as these will be the best of a variable bunch.

Buying hydrangeas

Hydrangeas are usually offered for sale in pots – either to use decoratively in the house, or to plant out in the garden. The main difference is that plants for interior décor are usually in flower when they are sold, whereas garden shrubs may or may not be in flower and may even be dormant, depending on the time of year.

In either case, try and pick a plant that is as healthy and well-grown as possible. Check for dead stems and moss growth around the top of the pots, which may indicate a lack of prior care. Hydrangeas tend to be trouble-free, but it still pays to keep a weather-eye out for pests or diseases.

By the time you buy your hydrangea, you will, ideally, have assessed your garden for climate, so use this knowledge to choose a species and cultivar which is likely to succeed.

HYDRANGEAS AS HOUSE PLANTS

With their classic good looks and long-lasting flowers, hydrangeas are increasingly popular as house plants – indeed, many of the earlier cultivars that were bred in France aspired to perfection as an interior decoration, and who are we to argue?

Nursery and supermarket shelves are stacked with pretty varieties, each one an instant, growing bouquet, but unlike other bunches of flowers, they are not transient and disposable. So, to get the very best out of your hydrangea, you must treat it kindly.

Site the plant in a part of your home that is cool, bright and airy, but does not get full sun. Avoid placing it near radiators or in very dry locations. Ensure the soil remains moist, but not wet and, since containerized plants rapidly run out of nutrients, feed it regularly during the growing season using a balanced fertilizer.

Most problems occur when the plant is kept too wet or dry. In damp conditions, fungal diseases such as botrytis may take hold; conversely, if the air and the soil are too dry, the plant may wilt, and dryness can also encourage red spider mite.

If desired, you can plant your "indoor" hydrangea outside, either in the ground or in a tub. But how well it does depends on care and breeding. Many glorious new plants are developed for the indoor- and cut-flower markets and their hardiness is not proven. Additionally, the plants that come foil-wrapped and in smart containers are hothouse-grown to be in magnificent bloom at the point of purchase, irrespective of the time of year.

But with the right cultivar and a moderate climate, the plants can be trained to cope with the outdoors. Put it outside in summer shade, ensuring that temperatures don't get too high or low, and that it doesn't dry out. Once it is acclimatized, plant it out and water well and regularly, or leave in a tub to bring indoors in winter.

Planting a hydrangea

Since hydrangeas are usually sold in pots, they can technically be planted into the garden at any time of the year – as long as the ground is not frozen and there isn't a heatwave. But if they have been forced on in a greenhouse, they must be acclimatized to your local conditions first.

Most plants are best planted out when dormant or growing slowly. Hydrangeas planted out in autumn, winter and early spring will have a chance to settle into their new location and get their roots down before the weather warms up, and this will help them to establish and reduce drought-stress. If you do choose to plant out a shrub in summer, you must water it frequently and well.

First make sure that the location is suitable to the species and cultivar, in terms of shade and moisture. Water the plant thoroughly, preferably by immersing the container in a bucket, then dig a hole approximately twice the size of the rootball.

On poor soils mix some additional organic matter into the earth that you have dug out, and a handful of your preferred slow-release fertilizer can also be beneficial.

Position the hydrangea so that the compost that was in the pot is level with the surrounding ground, then back-fill the loose soil and firm it with your foot. Take care not to bury the stem in the process or it may rot. Water well and, ideally, mulch the area around the plant to reduce competition and retain moisture.

GROWING IN POTS

Hydrangeas can be grown very successfully in containers and look great on a deck or either side of an entrance. It can also be a good way to cheat the local soil pH and get the flower colors you want. Indeed, in cold areas where the more tender *macrophylla* cultivars are marginal or otherwise impossible, container growing is the ideal way to enjoy them – displayed in a smart sleeve pot in summer, then lifted in its inner, utilitarian container, to be brought indoors to a frost-free place in winter.

The important thing is to pick a variety that doesn't grow too large, and don't skimp on the size of the pot you put it in. Use a good-quality soil-based compost that is designed for mature plants in permanent containers, and then you can tweak the pH if you want to (see page 28). Make sure that the container drains really well, but pay close attention to watering, particularly in summer, as dry conditions can inhibit flower production.

Depending on the ultimate size of the plant, you may need to repot your hydrangea every year or so, and, in any case, it is a good idea to remove some soil from the top of the pot and replace it with fresh on an annual basis. Feed and water regularly when in growth, prune as appropriate for the variety, and enjoy the show.

Feeding

All plants require good nutrition to perform at their best, so for the biggest flowers and most healthy foliage it is a good idea to add a little nourishment, at least annually. The amount and frequency of food that is required will depend to a certain degree on variety and soil.

Macrophylla types with their large leaves and often-heavy flowers tend to be pretty hungry, while others, including *Hydrangea petiolaris* and *H. paniculata*, are less so. If you have "perfect" rich, moist, free-draining soil, you may very well get away with applying a good mulch around the roots of the plant in late autumn or winter – something like spent compost or well-rotted manure, spread around the base of the plant, but not touching the stem, is ideal.

Plants on poorer soils should be mulched and also given a balanced feed a couple of times a year, when in growth. The additional nitrogen will help the stems and leaves grow fast and strong, but avoid feeding in late summer and autumn as new shoots produced after that point will be more susceptible to frost damage.

Many feeds and fertilizers are available, included pelleted manure, liquid feeds and slow-release fertilizers. Some of these are organic and some will be plant-based rather than containing synthesized chemicals and animal products. You can, therefore, feed according to your preferences; a little research can pay dividends.

Propagating hydrangeas

Multiplying hydrangeas is relatively easy and can be done by taking cuttings, by seed and, in some cases, by division.

Cuttings can be taken at almost any time of year. You get better results if you take your cutting material from a non-flowering stem.

Softwood cuttings You do this in spring and early summer, by cutting a new, flexible shoot, and trimming the bottom to just below a leaf node. You then remove the growing tip and most of the leaves and drop it into a premade hole in a pot of gritty, free-draining compost. (You can dip the base in rooting hormone first, if you wish, but I never do.)

Cover the top with a plastic bag to create a humid environment and place in a bright place out of direct sunlight. Roots should develop in 6–10 weeks, after which the plant can be potted up.

Semi-ripe cuttings These are taken in summer from the current season's growth, which should be hard at the bottom but still flexible at the tip.

Cut off a pencil-thick stem and trim it to about 4 inches long. Remove the soft tip and trim to just below a leaf node. Then remove most of the leaves and cut the remaining few in half to reduce water loss. Sink them into a pot of gritty compost and continue as for softwood cuttings.

Hardwood cuttings These are taken during autumn and winter, when the plant is dormant. Cut stems as per semi-ripe cuttings, but longer, at around 6–8 inches. Trim them neatly above and below a leaf node, and put them into the ground or in a large pot. A callus will form over winter, and will develop roots in spring.

Propagating from seed

Because hydrangeas often produce fertile flowers, particularly on lacecap varieties, you can collect seed in autumn to sow in spring. This works best for the species, since cultivars won't tend to come true, meaning that the new plants will have characteristics of both the seed parent and its pollen partner, which may have been a completely different-looking plant.

Division

In some cases, it is possible to dig up a multi-stemmed shrub and divide it into sections using a spade, saw or old bread knife. Then you can replant the sections where you want them, around the garden. Make sure that they are at the same depth as the original plant was (there is usually a line on the stem), firm the soil around the roots and water in well. Alternatively, you can pot them up to grow on or give away.

In addition, some varieties send up suckers, in which case you can just dig down around the periphery, cut off the new, rooted stem and pot it up to grow on.

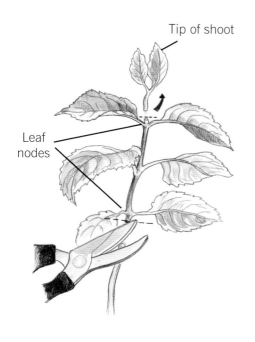

Tip of shoot

Leaf nodes

Cut stem below a leaf node

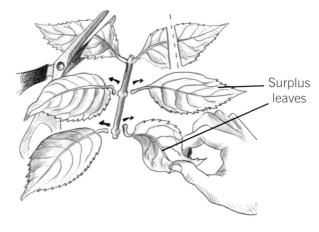

Surplus leaves

Remove leaves

Plunge into pot of compost

Bag

Cutting

Secure bag with string

Put in plastic bag

Pruning

Hydrangea flowers age attractively and the shrubs provide structure and detail in the winter garden, so in many ways there is no benefit to snipping off the spent blooms too early. In fact, some varieties do quite well with no pruning at all. In this sense, it is a plant that is ideally adapted for a modern gardener who may be busy, forgetful or simply can't quite be bothered.

Pruning serves a number of purposes. It allows plants to be shaped attractively and can help manage their size; it can also encourage the production of flowering stems, and manipulate the size and number of flowers that are produced. Finally, pruning can help maintain the health of a shrub, cutting out dead, damaged and diseased wood, and thinning it to let in more light and air, so it is less likely to develop problems, particularly with rots, in the first place.

There are four main approaches to pruning the genus *Hydrangea*, and which is best depends on which one you are growing.

Firstly, you can leave well alone. Varieties such as *H. hetrophylla*, *H. paniculata* and *H. heteromalla* can be left entirely unpruned, barring a quick snip every now and again if they need tidying. Likewise, the climbing hydrangeas need minimal intervention, although if you are growing one up the wall of a building, it is a good idea to trim it back from the windows and doors.

H. paniculata can be pruned in the same way as you would a buddleia, to create a small tree or multi-stemmed shrub. This can then be pollarded annually, to create a crown of flowers. The timing of this will also have an effect on when it flowers and the size of the panicle – light pruning in mid-winter, for example, resulting in larger, earlier flowers than hard pruning in mid-spring.

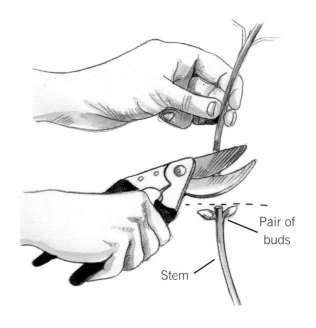

Pair of buds

Stem

For cultivars of *H. arborescens*, the pruning regimen can be tailored to your style. The lowest-impact approach is to simply snip off any remaining dead flowers in late winter – shears are entirely acceptable. To keep the plant compact or formal, you can thin stems by about a third, but some, including Annabelle, are notoriously floppy so it is wise to retain a supporting framework of old wood. The last option, for established bushes, is to cut back the entire thing to just above ground level. Although this seems brutal, as long as the hydrangea has a good root system, it will respond by producing vigorous new stems and larger-than-usual flower heads.

The varieties that cause most concern when it comes to pruning are the most tender members of the genus, *H. macrophylla* and, to a lesser extent, *H. serrata*.

These species develop flower buds on their stems towards the end of the year. These buds are carried through the winter and the flowers emerge the following summer.

The traditional way to prune these plants is to leave the old flower heads in place as a protection against frost and cut back to a pair of fat buds in early spring. The shrub can also be maintenance-pruned, where a number of the older stems are thinned out each year – removing the oldest wood to allow new wood to grow. This will let in light and air, and reduce competition, so the remaining flowers will be larger. A variation of this is to shear the bush lightly all over, reducing the stems to a pair of strong flower buds and trimming out any weak or damaged growth.

Cutting the shrub back further than this means that flowering is delayed while the plant rushes to develop flowering buds from lower down, but this can be useful if you are trying to get late flowers, for example for a late summer wedding, from a plant that would normally flower in early or midsummer. But, if cut back too hard, plants in a colder climate with a shorter growing season may not have enough time to produce flowering shoots at all. The exception to this is remontant or reblooming hydrangeas, which can be an asset in these areas (see page 206).

Finally, if you have a *macrophylla* or *serrata* hydrangea that is very tired and old, you can thin and reshape it by cutting it back as hard as you like, but this will be at the expense of the flower buds, so you can expect to get fewer blooms in the following season.

Pests and diseases

As plants go, hydrangeas really are remarkably trouble-free. Once established, they can roll on for years without flinching. They don't tend to require any sort of spraying regimen to keep them looking good, even in the most immaculately kept garden, so they are an excellent subject for low-maintenance and organic schemes.

There are still a few things that may, occasionally, rear their pestiferous heads, but good husbandry, vigilance and an eye to what it is that the plant prefers can help. Rather than battle a pest or medicate a disease, aim to have a plant which is healthy, well-grown and resilient, and manage conditions to discourage problems where possible.

Every country and region has its own horticultural challenges and what is a pest or significant disease in one place may be unheard of in another. The information below is designed to provide guidance and it is wise to refer to local experts for advice on your specific issues.

Pests

All gardens have their attendant pests, yet these creatures are often better thought of as wildlife in the wrong place. And, in a world where the natural environment is under threat and human impacts are substantial, it is not desirable to deploy chemical firepower.

In many cases, changing the location of the plant or doing something as simple as increasing or decreasing humidity can have a beneficial effect, and it may be essential to develop a degree of tolerance and encourage natural predators.

RED SPIDER MITE

This minute brown arthropod usually only becomes a pest in hot, dry conditions – which may, themselves, be causing the plant to struggle. Only just visible to the naked eye, large colonies can establish and the first thing that you notice is when the foliage becomes mottled and sick-looking. Closer inspection will reveal fine webs on the undersides of the leaves, in which you can just see moving mites.

The best thing to do is to remove and burn the worst-affected foliage, and spray the rest of the plant with a hose. Increasing humidity by watering and misting the plant helps, as the surface tension of the water droplets impedes the mites' ability to move.

Generally speaking, it is better not to spray against these pests as the chemicals may also harm other insects, including beneficial ones. Mites can be a particular problem on plants grown indoors, so if they persist, consider moving them to a cooler and damper location in the garden.

THRIPS

Known colloquially as thunder bugs, thrips are small, long-bodied, sap-sucking insects, which can become a problem in warm climates and when hydrangeas are grown under glass or as house plants. They can be kept in check by overhead irrigation and rain, however, when grown in cooler and outdoor locations; the populations are also kept in check by cold winters. By and large, thrips don't thrive in the woodland settings favored by hydrangeas, so the same things that will please the plant will also impede the pest.

Leaves attacked by thrips become dull green and then silver, and tiny spots of excrement may be visible. If you suspect thrips may be the cause of a sickly plant, shake a branch over a piece of white paper and inspect the results.

Generic pesticides will kill thrips, but they can be resistant and reservations about causing unintentional damage to other species apply. Increasing humidity by spraying the plant with water can help, and a biological control may be available for plants grown indoors.

HYDRANGEA SCALE INSECTS

Hydrangeas can occasionally suffer from scale insects, particularly hydrangea scale insect, which became established in the UK in the 1980s.

The scales are visible as limpet-like bumps on the stems and leaves, and some species, including hydrangea scale insect, produce white, oval egg masses in late spring and summer. Plants infested with scale insect grow less vigorously and can lose their leaves.

Heavily infested plants should be disposed of and replaced. Chemical control can be attempted in summer and organic options are available.

SLUGS AND SNAILS

Slugs and snails are a perennial problem and, on young bushes or smaller shrubs with soft fresh growth, they can cause significant damage. In

severe cases, they can graze away at the buds to the point where the plant dies.

There are many well-worn remedies. A wide barrier of grit may help and some molluscs can be trapped using beer or fruit. Organic slug pellets that don't poison pets or bio-accumulate in helpful slug-eating animals can be used *in extremis* but, in my experience, the most effective option is growing the plant on in a pot until it is big enough, and the leaves are tough enough, to resist attack.

APHIDS AND CAPSID BUGS

These are sap-sucking bugs which can accumulate at leaf tips and on fresh foliage, causing distortion and, in the case of capsid bugs, holes in the leaves. They can also secrete honeydew, which can result in an accumulation of black, sooty mold.

These bugs are easily visible and identifiable and, although various chemical treatments exist, blasting them off with water and encouraging natural predators such as ladybirds, lacewings and other beneficial insects is a gentler solution. Contact-acting soap-based sprays can help and 'environmentally friendly' products are sometimes recommended for both aphids and thrips.

VINE WEEVIL

Vine weevil can be an issue in potted plants, where the cream-colored, C-shaped larvae eat away at the roots. It is, however, very much less of an issue with plants in the open ground, where natural predators can reach them. Vigilance is sensible, as spotting them early will help, and they are best treated by watering on parasitic nematodes that will reduce the population levels.

DEER

In some places, browsing deer can be very destructive and, if you garden in one of these areas, you will already be creating barriers and fences to defend your plants.

As far as hydrangeas go, there are a number of practical solutions that can be deployed. The most sensible of these is probably to restrict your choice of plant to those species that will bounce back and flower, regardless of a good chewing, such as *Hydrangea paniculata*.

Alternatively, you can choose cultivars that are known to rebloom, or grow your plants in pots that can be moved inside as required.

Diseases

Hydrangeas are, in general, pretty disease-resistant. They don't tend to suffer from virus infection, so if the leaves become pale or blotchy, attention should be paid to feeding.

They can, however, occasionally suffer from a range of fungal issues. Powdery mildews can be a problem, particularly when the roots get dry, and a leaf-spot fungus called *Cercospora hydrangea* can be vexing in the USA. In the UK, meanwhile, honey fungus (genus *Armillaria*) is the commonest fungal problem.

In addition to this, hydrangeas can be afflicted with root rots if soil conditions are too warm and wet. As ever, good husbandry is key. Pay attention to the preferences of your plant and try to meet its needs, which includes making sure that the stems are not congested and that it is grown in a cool and airy spot.

GLOSSARY

AGM Award of Garden Merit, given by the Royal Horticultural Society, indicates that the plant is recommended by the society and will perform well in the garden.

Anther The pollen sac at the end of the stamen.

Apical bud/shoot The bud or shoot at the top of a stem.

Basic The chemical term for alkalinity; a basic substance has a pH higher than 7.

Bicolor Having two colors.

Callus A mass of undifferentiated cells that forms at the base of a cutting and eventually develops into roots.

Corymb A flat or flattish inflorescence where the outer flowers are carried on longer stalks then the inner ones, which brings them all to the same level.

Cultivar A cultivated form of the plant selected for its desirable characteristics.

Dwarf A smaller than usual cultivar of a plant.

Hybrid A genetic cross between two different species, genera or cultivars.

Inflorescence A group or cluster of flowers, and other organs such as stalks, reproductive parts, bracts and sepals, that make up the complete flowerhead of a plant.

Lateral bud/shoot A bud or shoot located on the side of the stem, often in the leaf axil where the leaf joins the stem.

pH The measure of acidity or alkalinity.

Potash The horticultural term for the element Potassium (K) in water-soluble form. The name comes from the original practice of collecting wood ashes in a container.

Remontant The quality of blooming more than once in a growing season.

RHS Abbreviation for the Royal Horticultural Society.

Species A population of individuals which have a high level of genetic similarity and which can interbred.

Sport A spontaneously arising mutation in part of a plant which can then be reproduced vegetatively as a new cultivar.

Stamen The pollen-producing (male) reproductive organ of a flower. It consists of a filament and anther.

Stigma, style The female reproductive structures of a plant.

Variety A classification of cultivated plants, below subspecies, where there are minor but distinctive and inheritable characteristics exhibited.

Vegetative propagation The process by which plants produce genetically identical new individuals, or clones.

INDEX

TRADE DESIGNATIONS

Throughout this book each hydrangea variety is referred to by its commercial name. In most cases the variety denomination in relation to worldwide Plant Breeders' Rights and any trade mark around the world relating to the commercial name has been omitted for ease of reading. The list below gives variety denominations and the trade mark status of the commercial names where applicable.

Hydrangea arborescens 'Incrediball'® = 'Blush'
Hydrangea arborescens 'Incrediball'® = 'Strong Annabelle' = 'Abetwo' (PBR)
Hydrangea arborescens 'Invincibelle Spirit II'® = NCHA2 = 'Pink Annabelle'
Hydrangea macrophylla 'Curly Sparkle Blue' (Curly® Series)
Hydrangea macrophylla 'Curly Sparkle Pink' (Curly® Series)
Hydrangea macrophylla 'Curly Sparkle Purple' (Curly® Series)
Hydrangea macrophylla 'Dark Angel Red' (PBR) (Black Diamonds® Series)
Hydrangea macrophylla 'Doppio Rosa' (PBR)
Hydrangea macrophylla Endless Summer® The Original = 'Bailmer' (Endless Summer® Series) (supprimer Bailmer) = 'Semperflorens'
Hydrangea macrophylla 'Expression' = 'Youmesix' (PBR) (You & Me® Series)
Hydrangea macrophylla 'Forever' = 'Youmeone'® = 'RIE01' (PBR) (You & Me® Series)
Hydrangea macrophylla 'French Cancan' (Rendez-vous Series)
Hydrangea macrophylla 'Glam Rock' = 'Pistachio'® ('Horwack') (PBR)
Hydrangea macrophylla Hovaria® 'Hopcorn' (PBR) (Hovaria® Series)
Hydrangea macrophylla Hovaria® 'Love You Kiss' (PBR) AGM (Hovaria® Series)
Hydrangea macrophylla 'Inspire' (You & Me® Series)
Hydrangea macrophylla 'Love' = 'H1917' (PBR) (You & Me® Series)
Hydrangea macrophylla 'Miss Saori' = 'H20-2' (PBR) (You & Me® Series)
Hydrangea macrophylla 'Passion' = 'Youmefour'® = 'RIE4' (PBR) (You & Me® Series)
Hydrangea macrophylla Rembrandt® 'Vibrant Verde' (Rembrandt® Series)
Hydrangea macrophylla 'Spike' (PBR) (Beautensia Series)
Hydrangea macrophylla Star Gazer = 'Kompeito' (PBR) (Double Delight™ Series)
Hydrangea macrophylla 'Together'® = 'Youmefive' (PBR) (You & Me® Series)
Hydrangea macrophylla 'Twist-n-Shout'® = 'Piihm-I' (Endless Summer® Series)
Hydrangea macrophylla 'Wedding Gown'® = 'Dancing Snow' (PBR)
Hydrangea paniculata 'Limelight' (PBR) AGM
Hydrangea paniculata 'Magical® Candle' 'Bokraflame'® (PBR) (Magical® Series)
Hydrangea paniculata 'Magical® Fire' = 'Bokraplume'® (PBR) (Magical® Series)
Hydrangea paniculata 'Magical® Flame' 'Bokratorch'® (PBR) = 'Mystical Flame' (PBR) (Magical® Series)
Hydrangea paniculata 'Polestar'® = 'Breg14'
Hydrangea paniculata 'Vanille Fraise'® 'Renhy'® (PBR)
Hydrangea paniculata 'Wim's Red'® = 'Fire and Ice' (PBR)

Hydrangea quercifolia 'Snowflake' = 'Brido' =
'Cloud Nine'
Hydrangea Runaway Bride® = 'Snow White' =
'USHYD0405' (PBR) (Garland Series)
Hydrangea serrata 'Tuff Stuff'® = 'Mak 20' (PBR)

AUTHOR'S ACKNOWLEDGMENTS

This book is dedicated to the memory of Margaret Duker.

I would like to thank my husband, Chris Wlaznik for his enduring positivity and encouragement and for offering wise words and perspective whenever required. And my children should be congratulated and appreciated for living with me through another book and developing excellent cooking and tea-making skills as they did so.

The hydrangea community has been hugely supportive throughout the writing and researching process. Particular thanks must go to Glyn and Gail Church, who I had the pleasure of coinciding with in person in a Somerset garden center, to Maurice Foster who opened my eyes to many less well-known species and Mal Condon, whose perspective on the American hydrangea world was invaluable. Thanks, too, to Robert and Corinne Mallet for helping iron out key details.

I am grateful to my friends and colleagues in the horticultural community for their wisdom and specific expertise, and particular mention must go to Guy Barter, Chris and Kimi Collins and also Andrew O'Brien for suggesting key tweaks, making me laugh and a shared enjoyment of language.

It is, as ever, a joy to work with Katie Cowan, Polly Powell and Izzy Holton at Pavilion, while Katie Hewett and Hilary Mandleberg have made the editorial process as painless as it ever can be, for which I am very appreciative. And thanks, too, to Somang Lee for her beautiful and informative illustrations, and Gail Jones whose elegant design pulls the whole thing together.

Finally, I must acknowledge the magnificent work of Georgianna Lane, whose eye for beauty is faultless and whose stamina, when it comes to pursuing plant varieties across countries and continents, is nothing short of astonishing! I know that everybody who picks up this book will be captivated by the fabulous photographs and inspired to enjoy hydrangeas in a whole new way. For me, there is a huge level of excitement when the pictures start to arrive, and creating this lovely book together has been thoroughly delightful.

PHOTOGRAPHER'S ACKNOWLEDGEMENTS

The glorious hydrangeas that appear in this volume were photographed in France, England and the USA at internationally recognized gardens and nurseries, field locations and in my own garden and studio.

Photographing Jardin Shamrock in Varengeville-sur-Mer, France, the French National Collection, was a special honor. Founded by Corinne Mallet in 1984, Jardin Shamrock is acknowledged as the world's premier hydrangea collection, with over 1,500 different varieties. It was a great pleasure and privilege to meet with her husband, Robert Mallet, who generously took the time to tour me around the garden, both in the summer and in early autumn, sharing his vast expertise and passion. Each visit ended with a delightful French country lunch among excellent company.

Kristin and David VanHoose of Hydrangeas Plus in Aurora, Oregon, likely the largest hydrangea nursery in North America, warmly welcomed us and gave full access to their extensive collection, even opening the nursery for us when they were away themselves, which enabled us to meet critical deadlines.

Roger and Fiona Butler of Signature Hydrangeas, a specialist division of Golden Hill Plants in Kent, England, kindly gave me the run of their wonderful collection where I photographed many of the lush, blooming varieties found in this book.

As with a previous title in this series, *Dahlias*, author Naomi Slade has been a cheery companion on the journey. Her vast knowledge and sparkling words bring this subject to life in a highly informative and thoroughly entertaining way that continually surprises and delights readers.

My husband, fellow photographer David Phillips, has once again been a vital member of the team, providing much-needed logistics, creative and technical support on location shoots. It's not too dramatic to say that without his presence, many of the images would not exist.

As ever, my gratitude to my family for their patience and understanding of my peripatetic lifestyle.

And finally, my heartfelt appreciation to Publisher Polly Powell, Publishing Director Katie Cowan, Designer Gail Jones, Project Manager Katie Hewett, Izzy Holton and the team at Pavilion Books for conceiving this book and selecting me to photograph it.

24 23 22 21 20   5 4 3 2 1

Published in the United States of America by
Gibbs Smith
PO Box 667
Layton, Utah 84041
1.800.835.4993 orders
www.gibbs-smith.com

ISBN 978-1-4236-5402-5
Library of Congress Control Number: 2019947341

Reproduction by Rival Colour Ltd, UK
Printed in Hong Kong

First published in the United Kingdom in 2020 by
Pavilion
43 Great Ormond Street
London
WC1N 3HZ

The Publisher has made every effort to include all Plant Breeders' Rights and
trade designations. Should any corrections be necessary we would be happy to
make the relevant adjustment in any future printings of this title.

PICTURE CREDIT
PAGE 43: RM Floral / Alamy Stock Photo

Naomi Slade is a journalist, author, designer and consultant. She works extensively within the gardening and lifestyle media as a writer and broadcaster and lectures on a range of specialist subjects. She is also an award-winning designer of show gardens and other exhibits, often with an environmental focus. A biologist by training, naturalist by inclination, and with a lifelong love of plants, her previous books include *Dahlias*, *The Plant Lover's Guide to Snowdrops* and *An Orchard Odyssey*. For more of her work visit her website naomislade.com or follow her on Instagram @naomisladegardening and Twitter @naomislade

Georgianna Lane is a leading floral, garden and travel photographer whose work has been widely published internationally in books, magazines, calendars and stationery. Her many other books include *Vintage Roses*, *Peonies* and *Dahlias*. Follow Georgianna on Instagram @georgiannalane or visit georgiannalane.com.